BUTTERFLIES & MOTHS

A KALEIDOSCOPE OF LIVING COLOR

With an introduction by Michael Tweedie

CRESCENT BOOKS

Contents

The photographs in this book were supplied by the following people: M. Bavestrelli (18–25, 28–30, 32–34, 36–49, 54–87, 105–115, 117–120, 124–136); I. Bucciarelli (7–13, 26, 50, 51, 88–90, 92–94, 102, 120–121); Bruce Coleman/Des Bartlett (116); D. Giussani (16, 52, 95, 123); S. Lombardi (1–6, 14, 15, 17, 27, 31, 35, 53, 103, 104); Michael Tweedie (96–100). The line drawings are by Joyce Bee.

Adapted from the Italian of Umberto Parenti

Printed in Italy by IGDA, Novara
Library of Congress Catalog card number: 72 – 84330

Butterflies are the only insects to have captured the human imagination in the way flowers and birds have done. Artists have exploited their brightness and symmetry for hundreds of years, especially those of China and Japan. All over the world people who either ignore other insects, or curse them when they devour crops, have told legends of butterflies, and identified them with spirits of the dead, messengers of good luck or, occasionally, of ill fortune. Country folk of earlier times led harsh laborious lives, and did not take much notice of butterflies; but they were always glad to see any sign of the end of winter. They welcomed the first Brimstone butterflies in the spring, and called the lovely insect 'butterfly' from the rich yellow colour of its wings. Now we use the name, rather unsuitably, to embrace all the familiar day-flying Lepidoptera. That is how the story goes, and it is as plausible as most other items of folklore.

Butterflies are indeed wonderful and beautiful creatures, and the more closely they are viewed the more beautiful and wonderful they seem. We marvel at the colours and patterns of their wings, but a microscope of only moderate power reveals orderly rows of tiny overlapping scales, which determine the patterns on the principle of a mosaic.

The extraordinary life history, starting from the egg, which hatches into a worm that eats and grows and then changes into something like an egg again, from which finally hatches the butterfly – all this is so familiar that our sense of wonder at it has become blunted. Nevertheless anyone who watches the emergence and development of a butterfly must find it hard to believe that the broad-patterned wings were packed into so small a space a few minutes ago; this is one of the spectacular miracles of natural history, and never fails to inspire joyous amazement when viewed for the first time, whether the spectator is an adult or a young child.

Butterflies are part of the heritage of wild life, and like most wild life are dwindling and retreating everywhere. Surely they are worth preserving, even if a few square miles of 'useless' forest and marsh must be spared in order to preserve them.

The pictures and text of this book are designed to give pleasure to those who already know something about butterflies and moths, and to provide a new source of enjoyment for the eyes and minds of those who are interested and would like to know more.

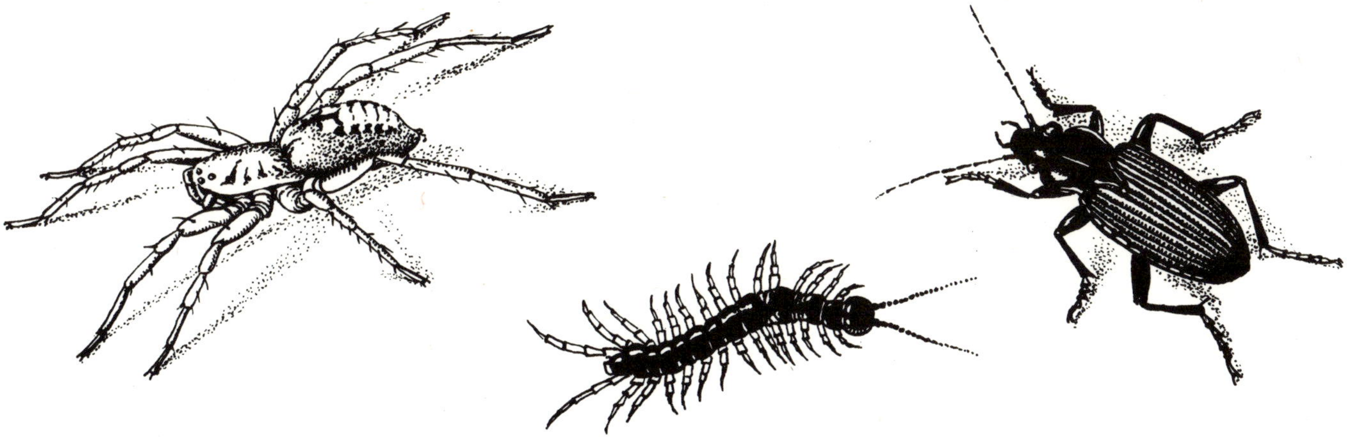

A spider, a centipede and a beetle; only the beetle is an insect.

The insect world

The butterflies and moths constitute an order called the Lepidoptera, of the class Insecta, the insects. To appreciate this book properly we must know just what an insect is.

Just as the Lepidoptera form a group within the Insecta, so do these make up a larger group, within one of the great divisions or phyla (phylum in the singular) of the animal kingdom, called the Arthropoda. It includes such creatures as lobsters, crabs, spiders, scorpions, centipedes, millipedes – and insects. The most important feature that all these arthropods have in common is the possession of a hard, jointed skeleton which encases the whole animal, head, body and limbs. This seems a strange idea at first because our own skeletons are internal, but in fact the jointed 'shell' of a crab's or a beetle's leg plays the same part as the bones in your leg or mine. The muscles are attached to the different segments in such a way that, by contracting, they cause them to pivot on their joints to make the complicated movements needed for walking or running. The difference is that the arthropod skeleton has the form of a rigid tube containing the muscles, while in a man, or any other vertebrate animal (that is, one possessing a backbone) it consists of rods of bone surrounded by the muscle. In insects the skeleton consists mainly of two substances; chitin, and the hard, rigid sclerotin.

The insects differ from the other arthropods, the crabs, spiders, centipedes and so on, in three important characteristics: when they are fully grown their bodies are divided into three distinct parts – the head, the thorax or chest, and the abdomen or hind-body. To the thorax are attached three pairs only of the jointed legs which are the hall-mark of the arthropods. Finally, most adult insects possess one or two pairs of wings, also attached to the thorax. No other arthropods ever have wings. The body of a butterfly conforms closely to this typical insect pattern, which is even easier to see in a wasp.

The head of an insect carries the mouth and main sense organs. The mouth is surrounded by a set of organs that either grasp or chew up the food, or are adapted for sucking it up in liquid form. The primitive condition, in which all the mouthparts are developed, is seen in a grasshopper or a cockroach, which has a pair of hard, sharp jaws for biting and chewing; but these work sideways, not up and down as ours do. There is an upper lip, the labrum, and a lower lip, or labium; and finally a pair of moveable maxillae just behind the jaws. The labium and maxillae have small jointed appendages like minute limbs; these are called palps and serve to manipulate the food and push it into the mouth.

The main sense organs on the head are the eyes and the antennae. Most insects have eyes of two kinds. On the top of the head are the ocelli, small simple eyes, incapable of forming an image. On each side of the head there is usually a large compound eye whose surface is divided into a number of tiny facets. These may number from several hundred to many thousands, according to the kind of insect. The large dragonflies which have very keen sight, have nearly 30,000 facets in each compound eye, and butterflies and moths have 12,000 to 17,000. Each facet is the outer end of a tube containing a lens constructed rather like a tiny telescope, and is called an ommatidium. The ommatidia converge towards the head and all meet on the surface of a light-sensitive retina. Each ommatidium conveys, not an image, but a spot of light, bright or dim according to the light coming from the direction in which the ommatidium is pointing. The image seen by the insect is therefore made up of brighter and darker spots on the same principle that is used for reproducing photographs in newspapers. Some insects, bees and butterflies among them, have colour vision, and their eyes are sensitive to shorter wave-lengths of light than ours, so that they see what we call 'ultra-violet' as a colour. They can also see the colours we do, with the probable exception of red, which is outside the limit of their vision in the same way that 'infra-red' is outside the limit of ours. It is interesting that many flowers have distinct patterns of ultra-violet coloration, visible to insects though not to us, but easily revealed by photography.

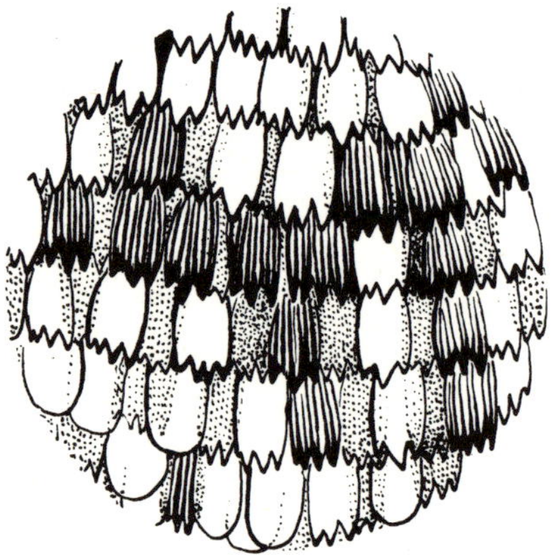

Part of a butterfly's wing, highly magnified to show the scales.

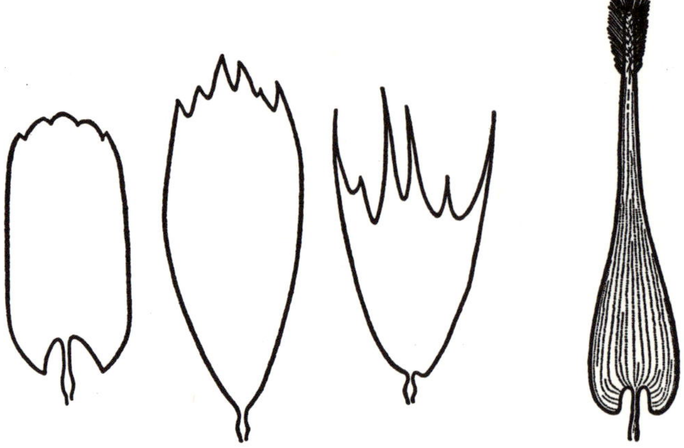

Different types of scales from the wings of butterflies.

The antennae of insects are mainly organs of smell. Instead of drawing air over an internal sensitive surface, as we do when we smell, insects catch molecules of scented substances on receptors, minute pits scattered over the surface of the antennae. In insects requiring a particularly keen sense of smell, the antennae may be elaborately branched to increase the number of receptors. Most male moths find the females of their own species by detecting a special scent given off by the female when she is in a sexually receptive condition. Some of the male Emperor moths (Saturniidae) will fly upwind to locate a female several miles away. In these moths the antennae of the males are formed like a feather, this feature being much less developed in the female.

By no means all insects have hearing organs, and when they are present they are not usually on the head. They may be situated near the junction of the thorax and abdomen, as in some kinds of moths; the crickets and long-horned grasshoppers have their 'ears' near the tips of their fore-legs.

An insect's head contains a small brain which interprets the messages received by the sense organs, but it is not essential to the working of the insect's body. A decapitated insect will continue to move about for a long time in an unco-ordinated sort of way, and may die from inability to eat and drink rather than from any disruption of its nervous system.

The thorax of an insect is essentially a box accommodating the muscles and nerves that activate its legs and wings; the alimentary canal, leading from the mouth to the stomach, runs through it. The abdomen, the hindmost part of an insect's body, contains the organs of digestion and reproduction, and sometimes accessory organs such as those producing the attractive scent in female moths, the light given off by glow-worms (which are really beetles) and the sting in bees and wasps.

Plate 15 (page 23) shows the large caterpillar of the Great Peacock moth, and along its side, just above the broken white line, a row of small, oval black-and-white spots can be clearly seen. These are the spiracles, the orifices through which the insect breathes. Our own respiration is effected by taking air into the lungs, where it comes into contact with the blood. This extracts the oxygen from the air and carries it round, through the arteries, to all parts of the body. Then the blood, depleted of its oxygen, is returned through the veins to the heart and lungs to repeat the cycle. Although insects have a rather simple system of blood circulation, it plays little part in distributing oxygen to the organs and tissues. Instead, air is admitted through the spiracles; there is usually a pair on each segment of the body, each of which opens into a tube that branches and re-branches to form a network of very fine tubules inside the body. In this way air is brought to the tissues and organs, but since the system is dependent on diffusion for its working, the air cannot be conveyed effectively for more than about half an inch from the spiracle. This is the chief reason why insects never grow bigger than a large mouse; no point inside the body can be more than half an inch from the surface because it would be starved of oxygen if it were.

Metamorphosis

The external skeleton of an insect, or any Arthropod, acts additionally as a coat of armour and, if watertight, prevents loss of water by evaporation; this is important for small animals as they would otherwise be quickly desiccated in dry conditions. But it has one great drawback: as it is not surrounded by living tissue, as bones are, it cannot grow. For this reason the outer covering of an insect has to be shed from time to time. When an insect sheds its outgrown outer skin a new one has already formed underneath. At first this is soft and stretches easily, but it soon hardens and further growth must wait for the next shedding.

The more simple and primitive insects differ from their parents, when they hatch, in size, but otherwise resemble them closely. They grow progressively by moulting their skins until they reach full size. This is also roughly true of such insects as bugs (Hemiptera) and grasshoppers, except that these are at first wingless, the wings appearing as

Butterfly eggs of three families (from left to right): Pieridae, Nymphalidae, Lycaenidae.

external buds and increasing in size and elaboration with each moult. Such insects are said to have incomplete metamorphosis.

The most advanced insects, including the butterflies and moths, go through a series of stages known as complete metamorphosis, which involves much more profound changes of form. These are well known: from the egg of a butterfly there hatches a minute larva or caterpillar which forthwith begins to feed, often making its first meal of its own eggshell and then turning to the foliage among which it finds itself. It feeds voraciously and grows rapidly, undergoing about five skin-changes (or ecdyses) in the course of its growth. When this is complete it hangs itself up in some way or spins a protective cocoon (the former usually if it is a butterfly larva) and becomes completely immobile. Its internal tissues and organs mostly break down into a sort of 'living soup' and their substance assembles itself again in the form of a butterfly, the wings being crumpled and folded and the legs packed tightly against the body. Only then is the last larval skin shed and the pupa or chrysalis revealed. The insect may remain in this stage, without the power to feed or move about, for two or three weeks, half a year or even several years. At the end of this time the shell of the pupa splits at the front end, and the butterfly or moth crawls out and climbs on to some kind of support from which it can hang while its wings expand and dry.

This type of development is shared by the beetles, the true flies and the bees, wasps and ants. It should be noted that all growth takes place while the insect is a larva. A small fly (or butterfly or beetle) cannot grow into a larger one.

Our final word about insects in general must concern their classification, which is based on the system invented by the Swedish naturalist Carolus Linnaeus in the eighteenth century. Under this system each species or 'kind' of animal is known by two Latin names, a specific name which identifies it precisely and a generic name which defines its closest relationship with other species. A group of species included under one generic name is called a genus (plural genera); the word species is the same in the singular and plural.

Species that have important features in common are grouped in genera and genera grouped in families, families in orders, orders in classes, classes in phyla and phyla in kingdoms, of which two are generally recognised, the Animal and Plant Kingdoms. Thus the Red Admiral butterfly is called *Vanessa atalanta* (its generic and specific names) and belongs to the family Nymphalidae, the order Lepidoptera, the class Insecta and the phylum Arthropoda. The fact that animals fall readily into categories expressing progressively wider degrees of relationship is, of course, simply an expression of the true relationships which have their origin in the process of evolution.

Butterflies and moths

It is unfortunate that there is no English word equivalent to the French '*papillon*' and the German '*falter*' to comprise the Lepidoptera as a whole. Instead we will use the scientific term, which is derived from the Greek words *lepis* (a scale) and *pteron* (a wing) and so describes the most distinctive character of this order of insects.

Anyone who has handled a butterfly will have noticed that the beautiful pattern on its wings is readily spoiled and blurred, and comes off on one's fingers in the form of a fine powder. A microscope or strong hard lens will reveal that this powder consists of tiny flattened scales, each one with a short stalk that fits into a socket on the surface of the wing. Examination of the undamaged wing shows that the scales lie in orderly rows, overlapping like tiles, and that each scale has a particular colour. The differently coloured scales thus make up the pattern, like a mosaic. No other insects have arrived at such an effective and versatile mode of colouring their wings, and the Lepidoptera have taken advantage of their scales to make their wings serve other purposes as well as flight, such as concealment or camouflage, warning coloration and recognition characters; we shall consider these in more detail later.

The other feature most characteristic of the

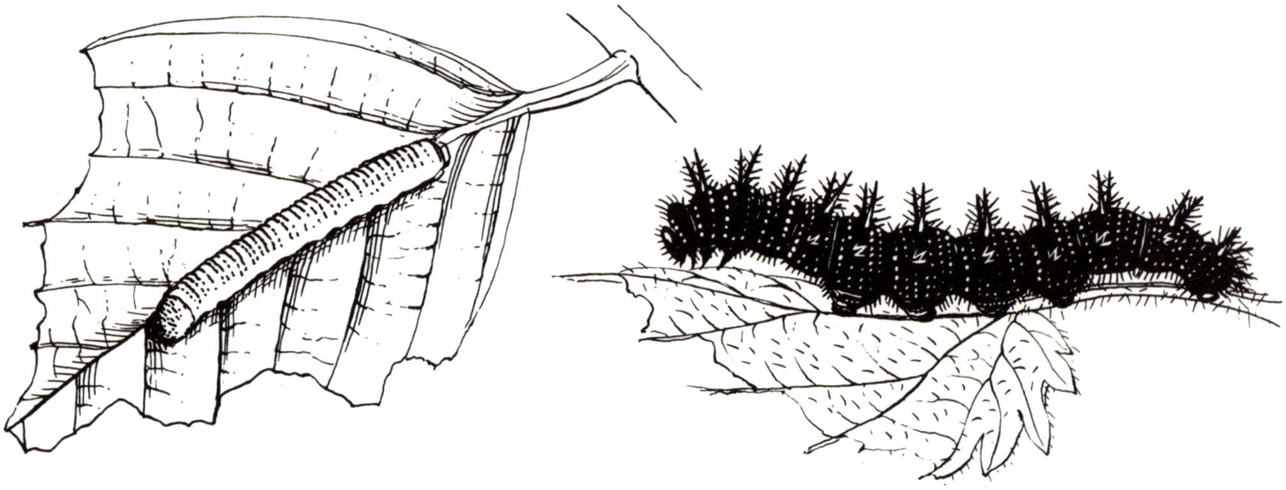

Butterfly larvae: Brimstone, Pieridae (left); Peacock, Nymphalidae (right).

Lepidoptera is the proboscis, or tongue. In our general introduction to the insects we described the mouthparts, including the mandibles or jaws and the pair of maxillae which lie just behind them. In butterflies and moths the mandibles have disappeared altogether and the maxillae have become profoundly changed, to form the right and left halves of a sucking tube which is normally carried coiled like the hairspring of a watch, under the head. When the insect feeds, the tongue is uncoiled and inserted into the nectar of a flower, or some other liquid food. The coiling and uncoiling of the tongue works rather like a steel tape which is curved from side to side, and stands out straight when it is pulled out of its case. In the butterfly's tongue the condition is reversed; its natural state is coiled, and it is extended and straightened by minute muscles all along its length, which cause it to curve from side to side. In some hawkmoths the tongue is extraordinarily long: for example, three and a half inches in the Convolvulus Hawkmoth – over half as long again as the head and body. Some tropical moths have the tip of the tongue stiff and spiny for piercing the rind of fruits; one Malayan species (*Calyptra eustrigata*), similarly equipped, has been found to be a bloodsucker, thrusting its dagger-like tongue through the hides of buffaloes and quite ready to take human blood if given the chance. The proboscis is, of course, only present in the imago or perfect insect. The larvae have well developed mandibles.

Distinctive features

These two characters, the scales, and the proboscis formed from the maxillae, are the main features that distinguish the Lepidoptera from other insects. Now we must deal with the frequently asked question: 'What is the difference between a moth and a butterfly?' It is in fact an unrealistic and unequal division of the order, by which one group of day-flying Lepidoptera is known as butterflies, and all the rest, including some groups really more distinct from each other than from the butterflies, are lumped together as moths. It has come about, of course, because the butterflies are mostly large and conspicuous and fly by day, while the moths include many very small species and are mostly nocturnal. The most obvious feature distinguishing the butterflies is the form of the antennae, which are knobbed at the end, those of the moths being variously formed but rarely knobbed. Most, but not all, butterflies rest with the wings held together over the back; only a few kinds of moths rest in this way. The situation is further complicated by the fact that the butterflies, as generally recognized, comprise two groups, the true butterflies or Papilionoidea, and the Skippers or Hesperiodea.

An even more unnatural division of the Lepidoptera is that which recognizes two groups, the Macrolepidoptera or butterflies and larger moths, and the Microlepidoptera or smaller moths. This is a device of collectors, made for their own convenience, because the very small moths are difficult to pin and mount and to identify. Although widely recognized this subdivision has no scientific standing at all.

The classification of moths is a complicated matter which we will not attempt to expound here. Butterflies are divided into a number of families, of which the following are represented among the pictures in this book:

Papilionidae: the Swallowtails, Apollos, Festoons.
Pieridae: Whites, Yellows, Orange-tips.
Nymphalidae: Emperors, White Admirals, Fritillaries, Vanessas and others.
Satyridae: Browns, Heaths, Ringlets and the Marbled Whites.
Lycaenidae: Blues, Coppers, Hairstreaks.
Hesperidae: includes the majority of the Skippers or Hesperioidea.

The Lepidoptera are the second largest order of insects, with over 140,000 known species; they are exceeded in diversity only by the Coleoptera, or beetles.

So much for the classification of the moths and butterflies; we will turn now to the more interesting subject of their way of life. We have already referred briefly to the life history of the Lepidoptera: the egg, the voracious and fast-growing larva, the limbless immobile pupa and the

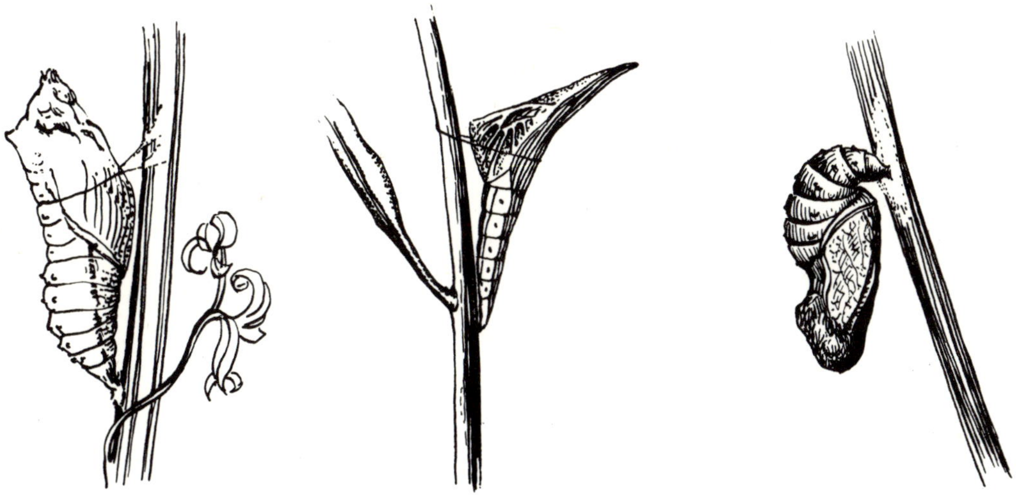

Butterfly pupae (from left to right): Swallowtail, Papilionidae; Orange-tip, Pieridae; Dark Green Fritillary, Nymphalidae.

active, vivid, often wonderfully beautiful imago or perfect insect.

The life cycle

The eggs of butterflies and moths are remarkably varied. Some are smooth and round, looking like tiny pearls, others have a more or less spherical shape with raised ridges or minute spines, others again are bottle-shaped or flattened like little buns, or some are curiously patterned with light and dark colours. The various families of butterflies tend to have eggs of definite types, which confirms a classification based mainly on the perfect insect. The eggs are usually laid by the female on the leaves of plants which will correctly nourish her caterpillar offspring; a Peacock butterfly will place her eggs on nettles, a Purple Emperor on sallow leaves, and a Small White on cabbages. On extrusion from the butterfly's body the eggs are covered with a sticky substance which firmly glues them to the surface of the leaf. One European species, the Map butterfly (*Araschnia Evana*) lays eggs on nettle leaves placing the eggs in chains or columns of a dozen or more, one on top of the other. Plate 3 shows a Leopard moth in the act of laying eggs on the bark of a tree. When they hatch the small larvae will burrow into the wood beneath them, eating their way through it.

The larvae or caterpillars of Lepidoptera are generally elongated, worm-like creatures with the three pairs of jointed legs, which mark them as insects, on the forepart of the body and usually five pairs of soft, thick, unjointed false legs or prolegs on the middle and hindmost body segments. These are provided with a curved row of minute hooks which enables the caterpillar to cling very tightly to a stem or a rough surface. In the family of moths called Geometridae the larvae have only the hindmost two pairs of prolegs developed, and progress by alternately advancing the fore part of the body, with the true legs and then the hind part, looping the body up as the hind part comes forward. They are often called 'looper caterpillars' or 'earth-measurers', referring to their movement.

Nearly all larvae of Lepidoptera feed on plants and the great majority on leaves. Usually the leaf is eaten from the edge inwards, but the larvae of some of the smallest moths burrow between the upper and lower 'skin' of a leaf, making a tunnel that appears as a light-coloured, usually winding ribbon, very narrow at its beginning, gradually broadening as the slowly tunnelling larva grows in size. Bramble or blackberry leaves are frequently marked by one of these leaf-miners, a minute moth called *Nepticula aurella*. A few caterpillars are partly or wholly carnivorous. That of the Dunbar moth (*Cosmia trapezina*) lives and feeds on leaves, but will seize and devour any other caterpillar it encounters. A collector who accidentally houses a Dunbar larva in one of his breeding cages will soon find it occupied by just one well-fed caterpillar.

One of the 'differences with exceptions' between butterflies and moths concerns the pupa or chrysalis. In most butterflies the larva hangs itself up in the open, either upside-down by the hind end or attached at the tail and by a girdle of silk round the middle. It then turns into a pupa, which is completely exposed and almost always protected by camouflage. The pupae of many kinds of Swallowtail are green when the larva suspends itself on a green background and brown if it is on a brown one. In the case of the Orange-tip, which spends most of the summer and all the winter as a pupa, it is green at first and gradually changes to brown, so that it is concealed among the lush vegetation of early summer and equally well when the leaves wither later in the year. Its curious pointed shape also helps to hide it among blades and stems of grass.

In nearly all moths the caterpillar spins a cocoon of silk underground, among litter on the ground, or sometimes on the trunk of a tree. In any case the pupa itself is wholly concealed and in no need of any adaptive form or coloration to protect it. This, no doubt, is the reason why most pupae have a uniform appearance like little black or brown bullets, and are very hard to identify.

A caterpillar spins silk by drawing a filament of sticky liquid from an organ called a spinneret on its lower lip. The filament immediately hardens and the caterpillar encases

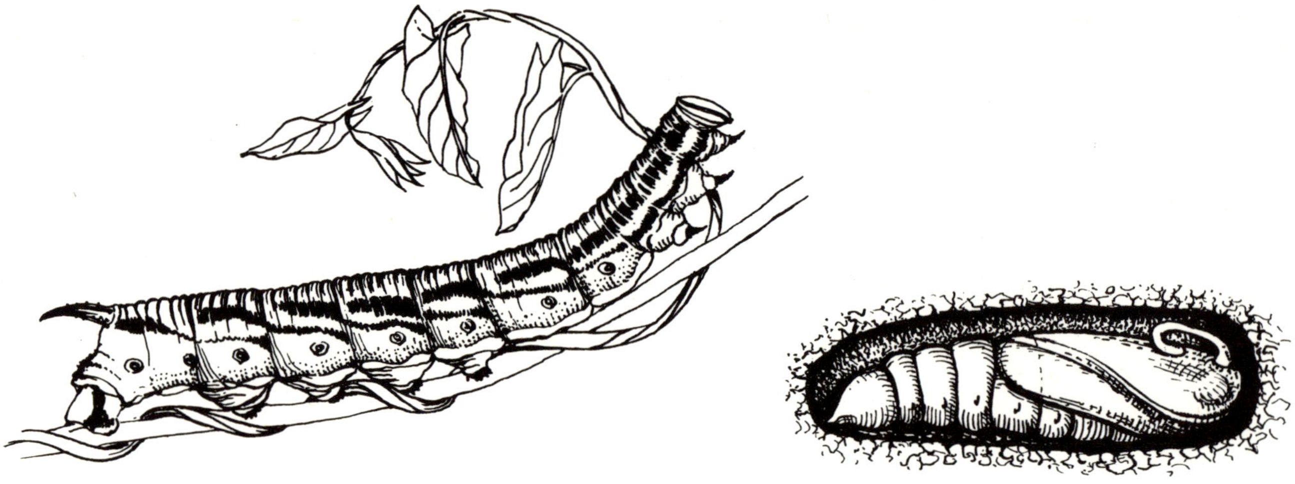

Larva and pupa of Convolvulus Hawkmoth; note the proboscis sheath on the pupa.

itself in a capsule or cocoon of silk by constantly moving its head and body. If the process is not interrupted the cocoon will consist of a single filament which, in the case of the domestic Silkmoth (*Bombyx mori*), is about a thousand feet long. This can be spun off and used to make thread for weaving the beautiful fabric called silk, a process discovered by the Chinese, according to tradition, about 4,000 years ago.

The European Puss moth (*Cerura vinula*) makes a cocoon on the bark of a poplar or willow tree, consisting of chewed-up fragments of wood stuck together with sticky silk. When this hardens it is almost invisible, and a knife is required to open it; it must be proof against the beaks of most birds. In the spring the moth softens the cocoon from the inside with a secretion of caustic potash, and so makes its way to the exterior.

An essential link in the chain of reproduction of nearly all insects is the meeting and mating of the adult male and female. In the Lepidoptera the males often hatch from the pupae earlier in the season than the females, and fly about looking for mates, at first in vain. Most of the females emit a scent when they hatch, which guides the males, as we noted when discussing the sense of smell located on the antennae. You will seldom catch a female moth that has not mated, so if a wild moth lays eggs after capture, you can be fairly sure that they will be fertile.

Butterflies, flying by day, are guided more by their sense of sight, and use this to find and recognize mates of their own species; this is one function of their bright, distinctive coloration. In many of them mating is preceded by a quite elaborate courtship. In the presence of a male, a female Small Tortoiseshell (*Aglais urticae*) will settle on a wall, or on the ground, and then turn this way and that, showing off her shining, tawny wings like a model displaying a gown at a fashion parade. She edges away as the male runs after her, sometimes allowing him to touch her wings with his antennae. The pair often fly away together, and return several times to the same place, before actually mating. Although butterflies do not find each other by their sense of smell, this often plays its part in the ritual of courtship. Many kinds of male butterflies have on their wings special scales, 'androconia', which are strongly scented. This scent is used to induce a receptive mood in the female; in the Danaid butterflies, such as the Monarch (*Danaus plexippus*) the androconia are actually brushed off the wings of the male and dusted over the female as he hovers over. In some butterflies the scent is clearly perceptible by the human sense of smell. That of the Wall butterfly (*Pararge megera*) is said to be like chocolate cream, and the male Green-veined White (*Pieris napi*) has a particularly strong scent like lemon verbena.

Methods of survival

Lepidoptera are beset by enemies in all their stages. Flying butterflies are pursued by birds, which they elude by their erratic, dodging flight; but some birds, such as flycatchers and bee-eaters, are very adept at catching them. Moths flying at night are hunted by bats, which are provided with an amazing and most efficient device for locating and capturing them. This is, of course, the well-known 'bat sonar' by which these animals emit ultrasonic sound pulses and instantaneously analyse their echoes from nearby objects, thereby avoiding obstacles in their flight and locating and identifying flying objects such as the moths on which they feed.

Most insects having special hearing organs, such as crickets and grasshoppers, and use them for listening to the calls of other members of their own species; but very few Lepidoptera make noises of any kind, and none are known to communicate by this means. However, many night-flying moths have hearing organs, some on the thorax, others at the base of the abdomen. It has been shown in recent years that these organs are sensitive only to ultrasonic vibrations lying within the range used for echo location by bats. Experiments have also demonstrated that moths respond to artificial ultrasonic sounds by dodging or dropping to the ground. Moths and bats have lived together for a good many millions of years, and it seems certain that these hearing organs have been evolved by moths purely as a

Convolvulus Hawkmoth hovering and feeding at a tobacco flower.

means of defence against these formidable enemies.

By far the most important enemies of Lepidoptera, however, are the parasitic insects which attack them in their early stages. These belong to two orders, the Hymenoptera, and the true flies or Diptera. The Hymenopterous parasites are wasps, which lay their eggs on or in the early stages of their prey. Usually the larva is the victim, but sometimes the pupa and even the eggs are victimized by minute wasps, a dozen or more of which may emerge from a single butterfly's egg.

The 'ichneumon' as the parasitic wasp is generally called, inserts its eggs into the caterpillar's body by means of a long organ, the ovipositor. The eggs hatch and the little grubs begin feeding on the tissues of their host, avoiding the vital organs, while the unfortunate caterpillar continues to feed and grow, and shows little sign of its affliction. Either just before or after it pupates, the ichneumon larvae come to full size and proceed to devour the whole body contents of their victim, putting an end to its life. They then pupate themselves, breaking out of the caterpillar's body, or, if it has turned to a pupa, remaining inside until the winged adult hatches or bites its way out through the shell. Some of the parasitic flies of the family Tachinidae lay large numbers of minute eggs on the leaves on which the caterpillars feed, and these are taken into its stomach, but the end result is just the same.

The Small White butterfly (*Pieris rapae*) is assailed by several of these parasites in its native territories in Europe and temperate Asia, and these are the chief factor in keeping its numbers to a reasonable level, though its larvae sometimes do quite serious damage to cabbage crops even in these regions. When the butterfly has been accidentally introduced to remote regions like New Zealand it has found itself in an environment devoid of its specific parasites, and has multiplied to the extent of becoming a very serious pest. The remedy in such cases is to breed large numbers of the parasites in caterpillars kept for the purpose in a country where the butterfly is native, and later release them in the territory which it has invaded. This has been done with very successful results in New Zealand and Tasmania, and the same method of biological control has been applied to a wide range of other insects; Lepidoptera are by no means the only victims of this sort of parasitism.

Moths and butterflies seem to have little defence against their parasites, and rely mainly on the principle that if a female lays several hundred eggs, a few of her offspring may escape death. Against enemies that hunt them by sight, on the other hand, they have evolved a variety of interesting protective devices. The great majority of these are adaptations for concealment. A majority of the caterpillars that feed among foliage are simply green, though many of them are shaded in a way that conceals the roundness of their bodies. Usually this means that they are darker above and paler below, so that light from the sky produces an effect of flatness. But the big, green larva of the Eyed Hawkmoth (*Smerinthus ocellatus*) is pale on its back and dark underneath. The right way up, it stands out sharply with accentuated shadows, but as it habitually rests clinging upside-down to a sallow twig, its countershading renders it almost invisible among the leaves.

Many larvae of the moth family Geometridae, the 'loopers' which we have already mentioned, are marked or coloured to look like small twigs. To perfect the resemblance they sit holding on only with the two posterior prolegs and holding the body straight and stiff at an angle to the twig on which they are resting.

Insects in disguise

We have mentioned the camouflage devices displayed by butterfly pupae. Many of the butterflies themselves are very effectively concealed in this way when they are at rest with closed wings. The European Commas and the American Anglewings (*Polygonia*) are good examples, but the most famous one is afforded by the Oriental Leaf butterflies (*Kallima*). When these butterflies are at rest the whole insect looks like a dead leaf. The outline of the fore- and hind-wings combined looks like a leaf, and a continuous dark line, simulating the mid-rib, runs across both wings

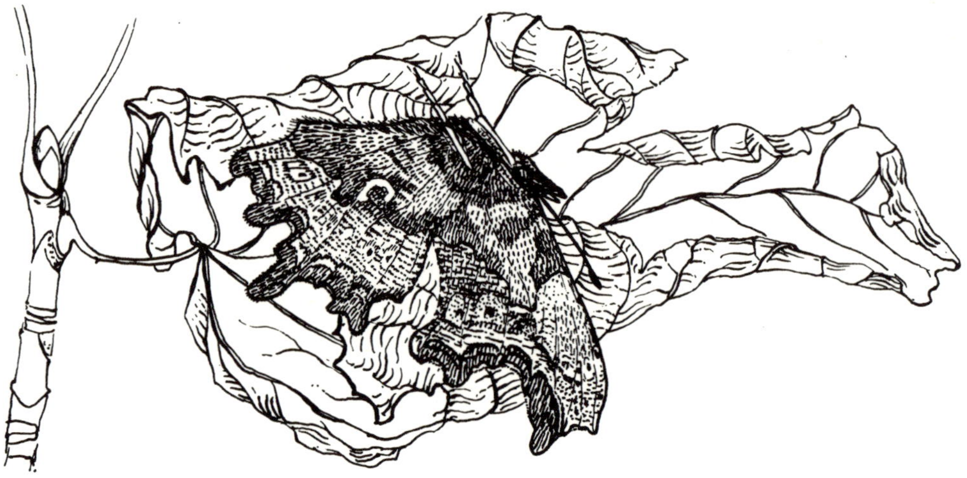

Comma butterfly hibernating, disguised as a dead leaf.

from the fore-wing tip to a short tail on the hind-wing, which represents the stalk. The upperside of the Leaf butterfly is blue and orange so that there is a startling explosion of bright colour when it flies and an even more puzzling complete disappearance of a bright colourful flying object when it settles.

There are moths also adapted for concealment by resemblance to leaves, and far more of them which are so marked and coloured that in their normal resting position, they blend with the bark of a tree, or a lichen-covered rock. Some of these, such as the Red and Yellow Underwings, have the hind-wings brightly coloured. When the moth is at rest the hind-wings are folded and completely covered by the fore-wings, which are coloured and marked for concealment. When the insect is alarmed by the too close approach of an enemy it suddenly flies and exposes its hind-wings. The effect is the same as that noted for the Indian Leaf butterfly, and this is an adaptation of such general occurrence among insects that it has received a name, 'flash coloration'.

Protected by poison

Some Lepidoptera do not require protection by concealment because they are inedible, being unpleasant to taste or even poisonous. Many of these feed on poisonous plants and retain the unwholesome and ill-tasting substances in their bodies, not only as larvae but throughout their lives. A good example of these are the butterflies of the family Danaidae, which includes the famous American Monarch butterfly (Plate 116). This species and its relatives feed on plants of the milkweed family (*Asclepias*) which contain heart poisons. In quantity the effect of these is fatal, but a bird that eats just one of these butterflies reacts by vomiting violently, no doubt suffering serious discomfort. This sets in almost immediately so that the bird associates its distress with the meal it has just taken. It has been shown, by experiments with aviary birds having no previous experience of eating Monarchs, that they will take one or two of them and suffer the consequences, and thereafter recognize their appearance and avoid them even if they are hungry. The Danaids have another remarkable adaptation associated with their poisonous nature. Their bodies are extremely tough and leathery and will stand up to the sort of rough treatment which would kill, at any rate cripple, a normal butterfly. Birds have been observed to catch Danaids, peck them and then, remembering or discovering that they are inedible, release them, and the insect just flies away. A palatable butterfly so endowed, would merely suffer a more protracted and painful death in the course of being eaten.

Two other groups of mainly tropical butterflies are protected in the same way. These are the black, white and red Swallowtails that feed on *Aristolochia* creepers, and the Pierid genus *Delias* of Southeast Asia and Australia, whose food plant is poisonous mistletoes of the genus *Loranthus.* Among European moths the Burnets (*Zygaena*, plates 4, 5) are both poisonous and extremely distasteful; a bird that pecks one releases it immediately, and wipes its beak, salivating copiously, to rid itself of the taste. They are said to derive their poison from cyanogenic (i.e. cyanide producing) strains of clover. The ragwort-feeding Cinnabar moth is another example; ragwort (*Senecio*) can have serious effects on cattle that are allowed to graze on it. The Tiger moths (family *Arctiidae*) of the temperate Old World and North America (plate 10) synthesize their own poison, and their caterpillars feed on a great variety of harmless plants. All Tiger moth larvae, however, are clothed in a coat of long bristles or hairs, which are enough to dissuade most birds from eating them.

Almost all the butterflies and moths which we have mentioned in the last three paragraphs are glaringly conspicuous in their coloration; red, yellow, black and white, sharply contrasted, being predominant colours. It is to their advantage to be immediately recognizable, and this advantage overrides that conferred by concealment. It is true that a few individuals are pecked or injured by inexperienced birds (not all have the additional defence of toughness enjoyed by the Danaid butterflies) but the

Courtship of Small Tortoiseshell butterflies.

general immunity to attack conferred on the species protected by this so-called 'warning coloration', combined with distasteful qualities, saves far more of their lives than the few that are sacrificed in the common cause.

The caterpillars, and often the pupae, are also protected by warning colours, and the former frequently feed crowded together and so render themselves additionally conspicuous. It must be remembered that larvae feeding on poisonous plants suffer far less destruction from the hazard of being accidentally chewed up by grazing animals, as wild animals generally know enough about their environment to avoid eating foliage that will disagree with them.

Butterfly mimics

The story of warning colours does not end with the poisonous and nasty-tasting moths and butterflies. Another mode of self-protection is developed from it. This is the remarkable phenomenon of protective mimicry, whereby edible species have, in the course of evolution, come to resemble inedible ones, gaining protection from their enemies by reason of the resemblance. Mimicry is particularly well demonstrated by butterflies; in North America, the abundant and distinctively coloured Monarch is not attacked by birds because if it is eaten it produces the same distressing symptoms as those caused by any Danaid butterfly. With it, but less commonly, flies another butterfly, the Viceroy (*Limenitis archippus*) a close relative of the White Admirals. However this insect is not black and white like a normal White Admiral; it is tawny brown with black streaks and white spots, just like a Monarch, and is of about the same size. In the course of the experiment in which inexperienced aviary birds attacked Monarch butterflies, suffered the unpleasant consequences and soon learned to leave them alone, Viceroy butterflies were also used. It was found that inexperienced birds attacked and ate them with relish and without any ill effects. On the other hand birds which had been ill after eating Monarch butterflies would not touch Viceroys either.

A more complicated example of mimicry is afforded by the African Mocker Swallowtail (*Papilio dardanus*). The male is a black and yellow butterfly with tails on the hind-wings. In some parts of its range the female looks like the male, but over most of tropical Africa the female Mocker Swallowtails lack tails on their wings and look nothing like Swallowtail butterflies. Nor do they all look the same: some closely resemble a black and white species called the Friar (*Amauris niavus*), others are almost perfect replicas of the African Monarch (*Danaus chrysippus*) which is black, white and tawny, very much like the American Monarch. Of course both the Friar and the African Monarch are poisonous. Here two more accompaniments of mimicry emerge. First, only the female Swallowtail is a mimic, never the male. Secondly the female is polymorphic: that is, she has more than one distinct form. Both of these conditions are frequently seen in mimicry associations, and can be attributed to the fact that the mimic must be, or at any rate appear to be, more rare than the distasteful model. If this were not so, birds would not thoroughly learn the lesson of associating the warning colour and pattern with inedibility, since they would encounter the palatable mimic too frequently. By confining the deception to one sex and by imitating more than one distasteful species the mimic will appear in smaller numbers than its models. Since the part played by the male in reproduction is quickly over, he can be regarded as relatively dispensable. The female must have a life of one or two weeks at least if she is to mature and lay all the fertilized eggs in her body. Nature is quite unsentimental in matters of this kind.

This type of mimicry was discovered in the last century by H. W. Bates among the butterflies of the Amazon Basin. Later a German naturalist, Mueller, working in the same region, described a second and rather more subtle variety: he found that unrelated groups of unpalatable butterflies had come to resemble each other. Since they are all protected by being inedible he suggested that, by all putting on the same uniform of warning coloration, they gave the birds fewer lessons to learn in avoiding them, thus reducing the wastage of lives expended in educating

A moth pursued by a bat, which finds its prey in the dark by echo-location.

inexperienced birds. These two forms of mimicry are accordingly called Batesian and Muellerian mimicry.

Hints for collectors

The easiest and most enjoyable way of learning to identify the butterflies and moths of a region is to make a collection of them, though this is hardly necessary in the case of British butterflies because they are so few, and well-illustrated textbooks so numerous. For the moths of any region, and for tropical butterflies, it is hardly possible to make any serious study of them without series of preserved specimens to compare. Detailed instructions for killing, mounting and storing Lepidoptera are to be found in almost every book that provides for their identification, and we shall not repeat them here, but only mention a few points that do not always receive sufficient emphasis.

The special pins, which are essential, are available from dealers in entomological equipment. They will supply the rest of the apparatus as well, such as setting or spreading boards, store boxes and cabinets, which must be very precisely made.

Label every specimen with at least the date and place of capture; it has no scientific value without such a label. Do as much breeding as you can, if possible from the egg; if you have the necessary skill, make drawings and photographs of the various stages. This is especially valuable in tropical countries, where many butterfly and moth life histories are still unknown. Try to identify food plants precisely.

Your collection should be a means of getting to know about insects in the first place, adding to scientific knowledge as soon as your own is sufficiently advanced. Butterflies are living creatures of great beauty, and the existence of many species is threatened by destruction of their habitat and by large-scale killing of them to make souvenirs and ornaments. Do all you can to discourage these activities, and remember that the collector who simply amasses great numbers of specimens for personal gratification is no better than the dealer who buys and sells them.

Index to colour plates

1

1, 2 *Small Ermine moths* These moths, belonging to the genus *Yponomeuta*, are placed among the so-called Microlepidoptera, the butterflies and larger moths being classed as Macrolepidoptera. The two terms have no significance in entomology but have been invented by collectors for their own convenience. Most of them confine their attention to the 'macros' because the small moths are difficult both to identify and to prepare for preservation.

Yponomeuta euonymella (1) is shown as an imago (perfect insect) and the larvae, or caterpillars (2) are those of *Yponomeuta padella*. The larvae of these moths always spin dense webs or nests among the leaves of their food plant, and a large number inhabit each nest. Those of *Y. padella* feed on plum, apple, hawthorn and blackthorn. Although the moths that are bred on these four food plants are indistinguishable, each plant supports what is called a 'biological race' of the species, and each race thrives only on its own food plant. The race confined to apple is named *Yponomeuta padella malinella*, and it may be abundant enough to be a pest in orchards. The moths are seen in July and August and the females lay their eggs on twigs. The caterpillars hatch in the autumn, but do not feed and spin their webs until late spring in the following year. They can be controlled by spraying or by cutting out the nests or burning them with a painter's blow-lamp. This is best done at night or in dull, cool weather, when the larvae are inside the nests.

Yponomeuta euonymella feeds as a larva on bird cherry and another species, *Y. cognatella*, on wild spindle and the cultivated form *Euonymus japonicus*, which it may damage severely.

2

3

4

3, 4 *Moths* Plate 3 is a much enlarged photograph of a moth in the act of laying its eggs. The larvae of this species, the Leopard moth (*Zeuzera pyrina*), feed on the living wood of various trees, including fruit trees, and the eggs are deposited on the bark. They are covered with a sticky fluid which firmly glues them to the substratum. The eggs of Lepidoptera are usually laid in this way, stuck down on to the surface of whatever type of food will correctly nourish the larvae. By far the majority of species glue their eggs to the leaves, buds or twigs of plants. The eggs are laid in groups, singly or in pairs. The female Lackey moth (*Malacosoma neustria*) affixes her eggs in a neat band or ring round a twig of blackthorn or apple, and they pass the winter in this situation. Some of the Eggar moths (*Lasiocampidae*) lay their eggs as they fly, scattering them over the ground below; their larvae are not very particular about what kind of plants they feed on. The Brown-tail and Gold-tail moths (*Euproctis chrysorrhoea* and *E. similis*) lay eggs in batches and cover them with a thatch of hair from the tuft which the moth bears at the end of the tail.

Eggs of Lepidoptera are very small and a female may lay from about fifty to several hundred, according to the species. The eggs may be smooth and rounded or have a characteristic form and sculpture. Those of the Pieridae or White butterflies (36–53) are shaped like tiny erect bottles. The eggs of the Nymphalidae (54–100) are usually rounded and ornamented with raised ridges.

The Leopard moth (3) is a European species that has been accidentally introduced into North America. As its larva feeds on the wood of trees, including fruit trees and those grown in city parks, the species is regarded as a pest, though not a very important one. Plate 4 shows a pair of Burnet moths (a species of *Zygaena*) mating. These brightly coloured, day-flying moths are specially abundant in the lands around the Mediterranean Sea. Although they are so distinctly coloured, separation of the species is often a difficult matter, and microscopic examination of their anatomy may be needed to be sure of the identity of a specimen.

5

6

5–10 *Moths* Plate 5 shows another Burnet moth, *Zygaena carniolica*. This is one of the most beautiful of the Burnets and is widely distributed on the mountains of southern Europe. Most species of *Zygaena* have the fore-wings ornamented with red spots on a metallic dark blue or green background, but in *carniolica* they are elegantly ringed with white.

The Plume moths (Pterophoridae) are mostly small insects and are classed with the Microlepidoptera, though some have a wing-span of five centimetres. They are peculiar in having the wings deeply cleft lengthways; the fore-wings into two parts, the hind-wings into three. Each part of the wing is fringed on each side and resembles a feather. The species illustrated (6) is the White Plume (*Pterophorus pentadactyla*), whose larva feeds on bindweed.

The larvae of butterflies and moths have three pairs of jointed legs on the anterior, or thoracic, segments of the body. These are the 'true' legs and are characteristic of all insects. They also have up to five pairs of blunt, unjointed 'false' legs or prolegs on the hinder part of the body. Most larvae have the full complement of five prolegs, and advance with an undulating motion. Those of the Geometrid moths have only the two hindmost prolegs developed (7) and walk by alternately straightening and arching the body. For this reason they are known as looper caterpillars or 'measuring worms'. The fantastic larva of the Lobster moth (*Stauropus fagi*, 8) has the second and third pair of true legs greatly elongated and only four pairs of prolegs.

On plates 9 and 10 are two moths of the family Arctiidae, the Ermines and Tigers. The species shown are the White Ermine (*Spilosoma lubricipeda*) and the Cream-spot Tiger (*Arctia villica*).

7 8

9 10

11

12

11–13 *Hawkmoths* This family, the Sphingidae, includes some of the largest and finest moths. Most of them are nocturnal; they have long narrow wings and thick, muscular bodies, and their flight is very powerful. Some of them are migrants and fly great distances; for example, several species which breed in North Africa are seen regularly in Britain. Plate 11 shows the Eyed Hawkmoth (*Smerinthus ocellata*) whose hind-wings are ornamented with beautiful eye-like markings. The Elephant Hawkmoth (12) is another common European species: its scientific name is *Deilephila elpenor*.

The larvae of the hawkmoths usually have a curved 'horn' at the end of the body, and this is clearly seen on that of the Elephant Hawk larva (13) which feeds in marshy areas on willowherb and bedstraw. The eye-like markings on the front part of its body give it an appearance rather like that of a snake, and it is possible that this is alarming to birds and so affords the caterpillar some protection. The front part, just behind the head, can be extended rather like an elephant's trunk, giving the species its name.

13

14

14, 15 *Moths* The family Saturniidae includes the largest moths, and indeed the largest of all Lepidoptera. Most of them are tropical, only a few species occur in Europe and one, the Emperor moth (*Saturnia pavonia*) in Britain. The biggest of all the European Lepidoptera is the Great Peacock moth (*Saturnia pyri*), and the imago and larva of this species are shown on this page; its wings may span as much as 16 centimetres. It is found in southern Europe, ranging as far north as Paris and Vienna, and in western Asia. The caterpillar feeds on blackthorn, apple and other fruit trees and is bright green with blue warts from which stiff bristles arise.

The great French entomologist Henri Fabre (1823–1915) made experiments to show that the males are attracted to the females over long distances by the scent which the female emits when she is in a sexually receptive state. The scent-perceiving organs of the male are located in the antennae, which are branched like a double-sided comb, the better to intercept air-borne molecules of scent. Males will fly upwind to find a female more than a mile away; at this distance the scent must be almost incredibly attenuated.

15

16

16, 17 *Skipper butterflies (Hesperiidae)* These are regarded as the most primitive of the butterflies and, in a way, as intermediate between butterflies and moths. They have the knobbed antennae so characteristic of butterflies, but their thick bodies and rapid, buzzing flight are more reminiscent of moths, and they rival the hawkmoths in their ratio of thoracic muscle to wing area. They reach their greatest development in the American tropics, where many large and brilliantly coloured species are found, but the Skippers of the temperate regions are mostly small butterflies patterned in shades of white, brown and blackish.

The Large Skipper (*Ochlodes venatus*, 16) ranges from Europe, including Britain, to Japan. Its caterpillar feeds on various kinds of grasses and pupates in a cocoon, another moth-like characteristic of many of the Skippers. The Large Grizzled Skipper (*Pyrgus alveus*, 17) is a mountain butterfly found in Europe, Asia and North Africa, mostly above 3,000 feet.

18 19

20 21

18–21 *Skippers and Blues (Lycaenidae)* The Safflower Skipper (*Pyrgus carthami*; upper side 18, underside 19) is widely distributed in southern Europe from the lowlands to as high as 5,000 feet, and flies over flowery banks and meadows.

The Large Blue (*Maculinea arion*, 20, 21) is chiefly remarkable for its curious life history. After hatching from the egg the larva feeds for a time on wild thyme flowers. After its third moult it leaves the food plant and wanders about until it is found by an ant of the genus *Myrmica*. The ant strokes a gland on the back of the larva, which responds by secreting a drop of sweet liquid. After drinking this offering the ant picks up the tiny caterpillar and carries it off to its nest. The larva remains there, feeding on the ant larvae and paying for its board and lodging with its sweet secretion. Having spent the winter in the ants' nest, it pupates in the late spring and emerges as a butterfly three weeks later, crawling out of the ants' nest into the open before expanding and drying its wings. If the young larva fails to encounter an ant it dies.

22

23

24

25

22–25 *Blues and Hairstreaks* The Common Blue (*Polyommatus icarus*, 22, 23) belongs to the same family, the Lycaenidae, as the Large Blue (20, 21) and the Purple Hairstreak (*Thecla quercus*, 24, 25). Most of the Lycaenidae are small butterflies and nearly all of them exhibit sexual dimorphism, that is the male and female are different in appearance. On the upper side of its wings, the male Common Blue is clear, bright blue with white fringes (22); the underside (23) is greyish-white with a pattern of black and orange spots. In the female the upper side is largely dark brown, with a little blue near the base of the wings and some orange crescents along their outer margins. The ground colour of the underside is also tipped with brown. The butterfly is common in flowery meadows, sand dunes and areas of rough grass.

The Purple Hairstreak is an insect of very different habits, for it is only found flying round the outer branches of oak trees, on which its larva feeds. Plate 24 shows the female upper side and 25 the underside. The upper side of the male is glossed all over with purple.

26

27

26, 27 *The Common Blue* Plate 26 shows a male sitting on a flower of the Birdsfoot Trefoil, a kind of vetch on which the larva feeds. In 27, seven of these butterflies are seen close together on the ground; they are indulging an appetite rather out of harmony with our usual ideas about butterflies. Although they feed on the nectar of flowers, these insects also gather at places where the ground is contaminated by droppings or urine of animals. Apparently they have a craving for the organic salts which are present in damp places of this kind, for when engaged in sucking up the moisture they are easily approached by the photographer or the collector.

30

31

28 *Swallowtail (Papilio machaon)* **29, 31** *Scarce Swallowtail (Iphiclides podalirius)* The butterflies shown on this and the opposite page all belong to the family Papilionidae. They range into every region where butterflies are found, except the Arctic, and are far more abundant and varied in the tropics than elsewhere. At the end of the book some tropical Swallowtails are shown (plates 124, 125, 127, 128, 131). The great Birdwing Swallowtails of the tropical Australian region include the largest butterflies in the world. Although a 'tail' projecting from each hind-wing is characteristic of these butterflies, it is not present in all of them. In some tropical Swallowtails there are even varieties of the same species which possess tails and others which lack them. Eggs laid by a tail-less female of the Asian *Papilio memnon* may develop into butterflies some of which are tailed and some not, and the converse is also true.

The larvae are very varied in colour and pattern, but all of them bear, in a transverse groove just behind the head, a Y-shaped organ called the osmaterium. This is usually orange or yellow and is normally hidden, but can be protruded if the larva is alarmed. It is connected with glands whose secretion, released when the organ is thrust out, has a strong odour. To humans it is pungent and unpleasant in some species but aromatic in others. There is little doubt that it acts as a defensive weapon, possibly against parasitic flies and 'ichneumon' wasps.

Some Swallowtail larvae have eye-like spots on the fore part of the body, rather like those on the caterpillar of the Elephant Hawkmoth (plate 13). The Spicebush Swallowtail of North America (*Papilio troilus*) shows these very well, and they may afford protection against birds by simulating a snake.

In all Swallowtails the chrysalis, or pupa, is attached to a stem or leaf by the tail, and by a girdle of silk round the thorax. It hangs head upwards in this position rather like a man working on a telephone pole. In some species, including the Common Swallowtail (*Papilio machaon*), the pupa may be green or brown in colour, and is usually green when attached to a leaf and brown if attached to a twig or a stone. It has been shown that the larva has a period of sensitivity to colour just before pupation, and this determines the pupa colour.

In Europe by far the most common species is *Papilio machaon*, usually known as the Swallowtail. It extends eastward to China and Japan and is the only one found in Britain. There it occurs as a distinct subspecies, *P. machaon britannicus* and is confined to marshes in the Broads of East Anglia. Elsewhere it is by no means confined to marshes, but flies over downs and any sort of open country. The caterpillar is green with orange-spotted black bands and feeds on various umbelliferous plants including wild carrot and fennel. A closely related species, *Papilio hospiton*, is found in Corsica and Sardinia and is entirely confined to these islands. The hind-wing 'tail' in *hospiton* is distinctly shorter than in *machaon*.

The Southern Swallowtail (*Papilio alexanor*) is confined in Europe to the area bordering on the Mediterranean, and is not uncommon in the foothills of the Alps. It is similar to *P. machaon*, differing in the black markings on fore- and hind-wings. The Scarce Swallowtail (*Iphiclides podalirius*, 29, 31) is pale yellow or whitish with dark bars and very long 'tails'. Its larva feeds on sloe and cultivated fruit trees. In Spain it is represented by a distinct subspecies, *I. podalirius feisthameli*.

In North America more than twenty species of Swallowtails are found, the most widespread being the Tiger Swallowtail (*Papilio glaucus*) and the Black Swallowtail (*P. polyxenus*). *P. machaon* extends to North America, where it is mainly subarctic and western.

32 33

34 35

30, 32–35 *Festoon and Apollo Butterflies* These are relatives of the Swallowtails and included in the same family. The caterpillars are similarly armed with an osmaterium.

The Festoons (genus *Zerynthia*) are a small group confined to southern Europe, North Africa and western Asia. They fly in rough stony places but do not range up to any great altitude in the mountains. They all have the beautiful intricate markings of the species shown here, the Southern Festoon (*Zerynthia polyxena*, 30), and are much prized by collectors. The larvae are dull coloured with red warts or spines and they all feed on species of Birthwort (*Aristolochia*), which are poisonous plants formerly used medicinally in childbirth. The Southern Festoon is distributed in southern Europe and western Asia Minor.

The Apollo butterflies (genus *Parnassius*) are highly characteristic of the mountains of Europe, Asia and western North America. They occur in greatest variety in central Asia, on and around the Himalayan massif, where they range up to great heights. Most of the species are more or less localized and some of the finest fly in extremely remote and inaccessible localities in Tibet, Western China and other regions even more difficult to visit today than they were earlier in this century. For this reason specimens of them are scarce and command high prices at the butterfly auctions which are held occasionally in one or other of the European capitals. Examples of these rarities are *Parnassius imperator*, the largest and finest species of all, from the Chinese-Tibetan border, and the very beautiful *P. charltonius autocrator* from the Himalayas. In far northern regions other species occur, usually in the lowlands where the climate and vegetation is similar to that of the high mountains farther south.

In North America several species of *Parnassius* are found in the mountainous regions of the far west from southern Canada to California, one of them being a subspecies of the far-ranging *P. phoebus*, a European race of which is shown on this page. It must be supposed that the evolutionary ancestors of the American Apollo butterflies entered the continent, via Siberia and Alaska, at a time when northern climates were less severe than at present, and when a land-bridge occupied the region of the Bering Strait.

Of the species shown here *Parnassius apollo*, the Apollo butterfly (32, 35), is the commonest and best known, occurring in all mountainous districts from Spain north to Scandinavia and east to central Asia. It flies at subalpine levels from 2,500 to 6,000 feet in the Alps and Pyrenees, lower in northern Europe. Numerous local races or subspecies have developed on the various isolated systems of mountains on which it lives. The caterpillar feeds on various kinds of stonecrop.

The Small Apollo (*Parnassius phoebus*, 33) is found in the Alps and in the Ural Mountains and Siberia and, as we have noted, in North America. It is rather smaller than the Apollo and its antennae are distinctly ringed with black and white. In the Alps it flies at higher levels than the larger and commoner species.

The Clouded Apollo (*Parnassius mnemosyne*, 34) extends from Europe to central Asia inhabiting mountains over most of its range, but flying in the lowlands in its northern localities. It differs conspicuously from the other two, and from most other species of *Parnassius*, in being without red markings. Its larval food plant is Corydalis.

A most peculiar feature of the butterflies is the appearance, after mating, of a brown, horny pouch at the end of the abdomen of the female. Its structure differs in the various species and affords a useful means of identifying them.

36

37

38

39

36–44 *Whites and Yellows, family Pieridae* Various shades of yellow, and white, are the characteristic colours of this family of butterflies, though some of the tropical species have brilliant red markings. Sexual dimorphism, or conspicuous difference between the sexes, is also a feature of many of them.

The Orange-tip (*Anthocharis cardamines*, 36, 37) is an outstanding example of this; it is hard to believe that the two butterflies at the top of this page belong to the same species. In the male, on the left, the tips of the wings are broadly marked with orange which in the female (37) is wholly lacking. When in flight she is difficult to distinguish from one of the butterflies collectively known as 'Whites'. A rare and remarkable type of 'freak' among butterflies is that known as a gynandromorph: that is, an individual in which one side of the body, including the wings, is entirely male, the other entirely female. In butterflies in which the sexes are similar this can go un-noticed unless the specimen is closely scrutinized, but there is no mistaking a gynandromorphous Orange-tip. Collectors pay high prices for such specimens.

The larva of the Orange-tip feeds on wayside plants of the family Cruciferae, including hedge mustard and lady's smock.

The Bath White (*Pontia daplidice*, 38) received its name in the eighteenth century, for the very whimsical reason that a young lady in the city of Bath had executed a piece of needlework in which the butterfly was represented. There is a specimen of the Bath White in a museum at Oxford in England that was captured and preserved in 1702; it must be one of the oldest butterflies in existence. The species has a wide distribution from southern Europe to India and Japan and in North Africa. In Britain it appears only occasionally as a migrant, but in 1945 large numbers reached the south of England and a summer generation bred there. Mignonette is the most usual food plant.

The arrangement of the chitinous supports or 'veins' in butterflies' wings plays a large part in the system by which they are classified. Usually the veins are concealed by the scales and special treatment is needed to reveal them. In the Black-veined White (*Aporia crataegi*, 39), the veins are clearly marked with black scales and the wings afford a beautiful demonstration of the venation of a Pierid butterfly. Its distribution is much the same as that of the Bath White, but it breeds considerably farther north. Until the 1920s it was a resident species in the south of England, its last stronghold there being in the county of Kent. Most Pierid larvae feed on low-growing herbaceous plants, but that of the Black-veined White lives on the leaves of Blackthorn and various fruit trees. The caterpillars live crowded together in silken webs and hibernate in them through the winter.

In the Green-veined White (*Pieris napi*, 40, 41) the sexes are distinct, the female (41) being more heavily clouded with dark scales and having a pair of dark spots on each fore-wing. This butterfly is found all over temperate Europe and Asia and also in North America, where it is known as the Mustard White. Many distinct subspecies exist, those of the New World being, as might be

40

41

42

43

44

expected, rather distinct from the Asian and European races. Experiments in cross-breeding subspecies from Europe and America have resulted in a high incidence of premature mortality (in the larvae and pupae), and of deformity in the butterflies. This genetic incompatibility suggests that the subspecies are evolving into full species.

In both this and the next two species there are two broods in the year, the butterflies appearing in May and August. Those of the spring brood are pale coloured like the male shown (40); summer males of the Green-veined White have a single distinct dark spot on each fore-wing, which is only faintly indicated in the spring form. The larva feeds on wild species of the Cruciferae plants.

The Small White or European Cabbage butterfly (*Pieris rapae*, 42, female) is the only species of butterfly which has a world-wide reputation as a pest of agriculture. It is a native of the temperate parts of Europe and Asia and its caterpillar feeds on the cultivated species of *Brassica*, cabbage, kale, cauliflower and the like. These plants have been introduced as food crops all over the world and the butterfly has followed them by accidental transportation. It first appeared in North America, near Quebec, about 1860 and quickly spread southward through the United States. It reached New Zealand in 1930 and southern Australia in 1939, whence it quickly crossed to Tasmania.

The Large White (*Pieris brassicae*) is also a pest of cabbage but has not spread in the same way as its smaller relative. Plate 43 shows the male, 44 the female.

45

46

47

48

49

45–51 *Whites and Yellows* The Clouded Yellows, genus *Colias* are butterflies of very wide distribution and occur all over Europe and temperate Asia, extending high in the mountains and even into the Arctic regions. They are well represented in North America and one of them, the Alfalfa butterfly (*Colias eurytheme*) has pest status because its larva feeds on alfalfa or lucerne, an important fodder crop. An interesting feature of the American Clouded Yellows is their extension, probably in some fairly recent geological period, all down the great Andean chain of mountains. Different species are found at different localities and some occur even in the cold stormy region of Tierra del Fuego in the extreme south. Almost all the species known are orange, yellow or white, but in central Asia an extraordinary blue 'clouded yellow' is found, *Colias sagartia*.

Two common European species are shown here, the Pale Clouded Yellow (*Colias hyale*, 45, male; 46, female) and the Clouded Yellow (*Colias crocea*, 47, male; 48, female). Both these butterflies are sometimes seen in Britain, the Clouded Yellow quite frequently; but they are summer immigrants, not resident species, as they cannot live through the British winter in any of their stages. The butterflies fly north from the Mediterranean region in May and June, crossing Europe and the English Channel, and lay their eggs in clover fields. From these eggs a generation of butterflies appears in August, but they all die, leaving no progeny, with the approach of winter. Female Clouded Yellows are normally yellow, as shown (48) but there is a form or variety named *helice* that is white instead of yellow. Males of this colour never occur, though the hereditary characteristic can be conveyed by males as well as females. Another species occasionally visits Britain in the same way; this is Berger's Clouded Yellow (*Colias australis*), which is so similar to the Pale Clouded Yellow that the two are difficult to distinguish.

The eggs of the Clouded Yellow are laid on lucerne or clover and are pale yellow at first, turning orange before they hatch. The mature caterpillar is dark green with minute black dots from which fine hairs arise, and there is a yellow or whitish line low down on each side. The pupa, like that of all Pierid butterflies, and also Swallowtails, is attached to a stem or leaf by the tail and by a girdle of silk round the middle. It is yellowish-green with fine dark markings and a beak-like projection on the head. The Pale Clouded Yellow is similar in its early stages, but the larva appears dark velvety green due to the presence of numerous minute black tubercles, each carrying a tiny hair. The pale line on each side is coloured white and yellow or orange. The larvae of nearly all the Clouded Yellow butterflies feed on leguminous plants, such as peas, clovers and vetches, but that of one species, the Moorland Clouded Yellow (*Colias palaeno*) is unusual in feeding on bilberry or whortleberry. It is found in the northern and Arctic regions of Europe, Asia and North America.

The Brimstone butterfly (*Gonepteryx rhamni*, 49–51) is a beautiful butterfly with the wings elegantly shaped and pointed. The male is deep yellow in colour, the female pale greenish-yellow. It is found all over temperate Europe and Asia as far as Japan, and its life history is unusual among Pierids in that the winter is passed in the adult or butterfly stage. In the autumn the butterflies seek out some dense evergreen cover, such as a thick growth of ivy, and hide themselves away, sleeping soundly through the period of frost and snow. On the earliest warm spring days they wake up and fly, and are among the first butterflies to be seen. It is believed that the bright yellow colour of the male is the origin of the English word 'butterfly', which has been extended to cover the rest of the day-flying Lepidoptera. 'Brimstone' is, of course, an old word for sulphur and also refers to the colour of the male. The caterpillar feeds only on the leaves of the small trees of the genus *Rhamnus* or Buckthorn.

The Brimstone is a common butterfly in Britain; in southern Europe another species, the Cleopatra butterfly (*Gonepteryx cleopatra*) is often seen. It resembles the Brimstone but the central part of the fore-wing is deep orange in the male. The female has only a faint orange streak on the underside of the fore-wing and cannot be distinguished in flight from *G. rhamni*. The larva also feeds on buckthorn, and the butterfly hibernates through the winter in the cooler parts of its range.

50

51

52

53

52, 53 *Pale Clouded Yellow and Orange-tip* Both these species have been shown in photographs of pinned and mounted specimens, and such pictures are necessary in order to display their characteristics to best advantage.

Photographs of living butterflies, taken in the wild, have the advantage of showing their natural posture, and also they do not involve catching and killing the insects. In heavily populated and cultivated regions butterflies are already far less abundant than they were only a few years ago, and many species require protection if they are not to vanish altogether. Collecting butterflies simply to fill cabinets with trophies is an outmoded pastime, and a collection of photographs of the kind displayed on this page will bring a naturalist far more credit than rows and rows of dead specimens. Not only does photography spare the insects' lives, it calls for far more patience and skill than catching them with a net. A camera adapted for close-up work is needed, and on dull days, when butterflies are easier to approach than in full sunlight, electronic flash gives the photographer a great advantage.

54 55

56 57

54–57 *Purple Emperors* The two upper plates show the Purple Emperor (*Apatura iris*), and the two lower the Lesser Purple Emperor (*Apatura ilia*). In both cases the male is shown, upper side on the left and underside on the right. In spite of their names there is little difference in size between the two. The females of both are rather larger and lack the beautiful purple gloss on the fore-wings.

These two species serve to introduce the family Nymphalidae which includes a great variety of butterflies, many of which are large and robust with powerful flight. In all of them the fore-legs are reduced in size and not used for walking, so that these butterflies are functionally quadrupeds. The terminal joints of the stunted fore-legs are hairy, and the Nymphalids are often called 'brush-footed butterflies'.

In both the species illustrated the males have a purple gloss on the fore-wings. This is not produced by pigment but is an example of what is known as structural coloration. It is due to extremely minute, ultra-microscopic, irregularities on the scales, which have the effect of breaking the light up into its different wave-lengths; colours of this kind usually change with the direction from which the light strikes them, or the angle from which they are viewed, so that one side of the butterfly may display bright colours which do not appear at the same time on the other side. The iridescent colours of a film of oil on water are produced in the same way.

The Purple Emperor is confined to forests or extensive woodlands, especially oak woods, and the males fly round the tops of the tallest trees. This preference for oak woods is rather mysterious as the caterpillars do not feed on the leaves of oak, but of willow and sallow, especially goat willow, *Salix caprea*. This species ranges from Europe across Asia to Japan and is found in some of the larger woodlands in southern England. The Lesser Purple Emperor has a similar range but is absent from Britain. It is found in open woodland where poplar and aspen grow, these being the food plants of the larva.

58

59

60

58–60 *White Admirals* The Poplar Admiral (*Limenitis populi*, 58) is found in continental Europe, including Denmark and Sweden, but is absent from Italy, Spain and Portugal and does not occur in Britain. It differs from the other White Admirals in having submarginal orange crescents on the wings and no large white spots or bars. It is found mainly in rather humid localities in the lowlands, where the food plants of its larva, aspen and poplar, grow freely. The White Admiral (*Limenitis camilla*, 59) has a beautifully marked underside (60). It is a smaller butterfly than the Poplar Admiral and is the only member of the genus found in Britain, where it is confined to woodlands in southern England. In Europe it is widespread but does not extend into Spain. The larva feeds on honeysuckle and, when young, devours the leaves in a peculiar way, eating them back from the tip and leaving the mid-rib.

One other species of White Admiral is found in Europe; this is the Southern White Admiral (*Limenitis reducta*), which is rather similar to the White Admiral and is the only species of *Limenitis* found in Spain. In North America several species of *Limenitis* occur. One is called the White Admiral (*L. arthemis*) and another is the well known Viceroy (*L. archippus*) which resembles the Monarch butterfly (116) in appearance.

61 62

63 64

61–64 *Fritillaries* These butterflies have their name in common with the flowering plant *Fritillaria*, and the derivation in both cases is from the Latin word for a checker-board; the reference, in the butterflies, being to the checkered markings on the upper surface of the wings. They are characteristic of temperate northern regions, both of Eurasia and of North America.

There are two main groups of Fritillaries. In one there are silvery or pearly markings on the underside of the wings; in the other group these are lacking, and they are named 'Checkerspot' in the United States. The Queen of Spain Fritillary (*Issoria lathonia*, 61, 62) is widespread in Europe and North Africa but known in Britain only as a rare migratory visitor. Its underside, shown in the right-hand upper figure, is more richly adorned with metallic spots than any other European species. Like most of the true Fritillaries its caterpillar feeds on leaves of violet. The Spotted Fritillary (*Melitaea didyma*, 63, 64) belongs to the 'Checkerspot' group, its underside being devoid of metallic spots, though beautifully marked as the right-hand figure shows. It is widespread in Europe and central Asia and extremely variable; a southern 'Mediterranean' form, a northern 'European' form and an 'Alpine' form have been distinguished, but intermediates between them occur. The butterfly flies in flowery meadows and its larva feeds on toadflax and plantains.

65

66

67

65–67 *Fritillaries* Cynthia's Fritillary (*Euphydryas cynthia*, 65, upper side; 66, underside) has white markings on the upper side of the wings which are quite unusual for a Fritillary, and are only seen in the male. The female is orange-brown with blackish checkered markings. This is one of the most celebrated European butterflies, both on account of its unusual coloration and its very restricted range: it lives only in the Alps and the mountains of Bulgaria at altitudes of 6,000 to 8,000 feet. Those found at the highest levels are generally small but brightly marked. The butterflies are attracted by juniper bushes, but the larvae, which are very hairy, live on various low-growing plants including plantain and lady's mantle.

The Small Pearl-bordered Fritillary (*Clossiana selene*, 67) has an extremely wide range across Europe and Asia, including Britain, and its subspecies, *C. selene myrina*, is found in North America. It flies at the margins of woods and in clearings, from sea level to 6,000 feet.

68

69

70

71

68–71 *Fritillaries* The Dark Green Fritillary (*Mesoacidalia aglaja*, 68, 69) extends over the whole of Europe, including the British Isles, and across Asia to China and Japan. It is one of the larger Fritillaries and, as the top right-hand figure shows, the silvery spots on the hind-wing underside are well developed. The dull green background of these spots gives the butterfly its English name. It flies in open country such as heaths, meadows and coastal sand dunes, and in the mountains extends upwards to the tree-line. The caterpillar is velvety black with grey segmental rings, a pale stripe along the back and numerous branched black spines. It feeds on violet.

The Silver-washed Fritillary (*Argynnis paphia*, 70, 71) is a very fine butterfly, even larger than the Dark Green, and has silvery stripes on the hind-wing underside in place of the well defined spots seen in most of the other species. The sexes are quite distinct; the black streaks on the fore-wing of the male (70) are lacking in the female. There is a variety of the female in which the ground colour of the wings is olive-green instead of fulvous. It is called *valesina* and never appears in the male. The case is similar to that of the Clouded Yellow with its female variety *helice*, see 47, 48. The Silver-washed Fritillary flies in clearings in woods and is fond of visiting the flowers of bramble.

72 73

74 75

72–79 *Vanessid butterflies* These are the most familiar members of the family Nymphalidae, the brush-footed butterflies, and are characteristic of the northern temperate regions of Eurasia and America. The caterpillars of some of the commonest feed on stinging nettles; indeed, if we exterminate nettles, some of the most beautiful butterflies of our gardens will disappear.

The Painted Lady (*Vanessa cardui*, 72, underside 73) has the distinction of being the most widely distributed of all butterflies. It is represented by a distinct subspecies, *kershawi*, in Australia, in which the black spots on the hind-wings are centred with blue. On all the other continents it is distributed without regional variation. The reason for this is that the urge to migrate is more powerful and persistent in the Painted Lady than in any other butterfly. Its populations are therefore subject to constant mixing which prevents the formation of local races. The main breeding ground of the European Painted Ladies is in North Africa, where the chrysalids hatch in thousands in dry grassy places and sand dunes, and then start a massed flight of butterflies northwards to the Mediterranean.

In Britain and northern Europe the migrants arrive in the spring, spreading all over the country, and some even press on to Iceland and places within the Arctic Circle. Where the climate is suitable an autumn generation of butterflies is produced from eggs laid by the early migrants, and these are reinforced by further migration, but they cannot survive the winter in Britain or anywhere far north of their breeding grounds. They are seen in the northern countries every year, but the butterflies are much more abundant in some years than others. The larva feeds mainly on thistles, and this is indicated by the Latin name *cardui* and also by the German name 'Distelfalter'. In France it is called 'La Belle Dame', a similar name to the English.

The Red Admiral (*Vanessa atalanta*, 74, 75) has not the enormously wide distribution of the Painted Lady, but it is found in Europe and Asia as far east as Iran and also in North America, extending south to Guatemala. It is a less compulsive migrant than the Painted Lady, but its status in Britain is similar. Red Admirals fly north and cross the English Channel in the spring, and the butterflies seen on michaelmas daisies and fallen fruit in the late summer are their progeny. These Red Admirals are unable to winter in Britain, and there is some evidence that they fly southward, but it is not known if any of them ever reach their breeding grounds in southern Europe and North Africa.

Experiments have been made with Red Admiral butterflies which explain the use of the stunted forelegs. In this species, at any rate, they are organs of taste. By touching the liquid with its 'finger-tips' a Red Admiral can distinguish between pure water and a sugar solution $\frac{1}{200}$th of the strength needed to register on the human tongue. It is possible that the forelegs serve this purpose in all the

76

77

78

79

brush-footed butterflies. The larva of the Red Admiral feeds on nettle.

Two kinds of Comma butterflies are found in Europe. The Southern Comma (*Polygonia egea*, 76, 77) is a butterfly of the south-eastern region of the continent, absent from most of France and Spain and quite unknown in Britain. It flies in dry rocky valleys and its larva feeds mainly on pellitory. The Comma (*Polygonia c-album*, 78, 79) takes its name from the white comma-like mark on the underside of the hind-wing, and the Latin name means 'white-c'. It has a far wider distribution than the other species, extending from most of Europe, including southern Britain, across Asia to Japan.

The non-migratory Vanessid butterflies winter in the imago or butterfly stage, and the Comma hibernates in the open among herbage or in a hedgerow. When its wings are closed their very irregular outline and intricate brown pattern produce a remarkable resemblance to a withered leaf; even the white 'comma' plays its part, looking like a crack with the light shining through it. Small birds are continually hunting for food throughout the winter in just the situations where the Comma hibernates, clinging motionless to a twig. Its elaborate camouflage serves to protect it from their sharp searching eyes. The caterpillar feeds on nettle, hop and various trees, especially elm.

Quite a number of species of Polygonia live in North America, where they are called commas or angle-wings.

80 81

82 83

80–87 *Vanessid butterflies* The Camberwell Beauty (*Nymphalis antiopa*, 80, underside 81) is so called in Britain because in 1748 some specimens were taken at the village of Camberwell. This rather fanciful name serves to remind us that Camberwell, now well inside the urban sprawl of London, was at that time a pleasant rural area with fields and trees and clean, flowing streams. It is also a reminder that butterfly collecting was already established as a pastime in England in those days. The capture of these specimens was recorded in 'The Aurelian', a book written and illustrated by Moses Harris and published in 1766. It includes descriptions and figures of 33 British butterflies, about half the number that were to be found over the whole of Great Britain. The English names in 'The Aurelian' were the first that came into general use, and most of them have been retained, including 'Camberwell Beauty', but this butterfly was given an alternative name: 'Grand Surprise'. In those days, as now, it was a great rarity in the British Isles.

Elsewhere it has a very wide distribution across temperate Europe and Asia and in North America, where it is known as the Mourning Cloak, a translation of its German name Trauermantel. This is a reference to the dark, pale-edged coloration of the underside of the wings (81). Its occasional appearance in Britain may be due to migration, but it has been suggested that the pupae or hibernating butterflies are sometimes brought over in timber-carrying ships from

84

85

86

87

Scandinavia, where the butterfly is common. It seems to need a continental climate with hot summers and hard winters to complete its life cycle and period of hibernation, and cannot thrive in the wet, equable Atlantic climate of the British Isles. The larvae feed on the leaves of various trees, including birch and willow, and live crowded together in webs for most of their lives. This is a handsome caterpillar, black spotted with yellow and red.

The Large Tortoiseshell (*Nymphalis polychloros*, 82, underside 83) is found in Europe and eastward into Asia as far as the Himalayas, and also in North Africa. In Britain it was fairly common in southern and midland England until 50 to 60 years ago, when it rapidly declined in numbers and is now restricted to a few localities in Essex and Suffolk. Its larva usually feeds on elm, but is found on birch and various willows as well.

The beautiful and unmistakable Peacock Butterfly (*Inachis io*, 84, underside 85) ranges all across Eurasia but does not extend to North America. Like all the butterflies on these two pages it hibernates as a perfect insect, and often shelters for its winter sleep in outhouses or attics. In England an experiment has been made in domesticating the Peacock. Females are caught in spring and confined with nettles, grown in pots, and they lay eggs on them. The caterpillars are reared in the usual way and the butterflies emerge in July and August. At first they are kept for a week or so in a cool greenhouse in which they can fly freely and feed from bunches of flowers in water, and from saucers of honey and water. In these conditions they quickly become tame and will learn to come to one's hand to feed from a spoon. Next they are released in the garden and will stay there if there are plenty of flowers and if hand feeding with honey and water is continued. The greenhouse door is left open and they will come into it to sleep at night. In the autumn as many as possible are caught and put into large muslin cages which are kept indoors, and feeding is continued until the weather becomes cold. The cages are then put into a permanently dark attic or outhouse where they will be cold but not exposed to frost. More will survive through the winter if they are brought into a warm room and fed every three weeks or so. In the spring the cages are put in the greenhouse and the butterflies released and fed daily. As soon as flowers are available they can be allowed to fly in the garden and should again be lured and tamed by artificial feeding. Under these conditions a Peacock has been known to live, as a butterfly, for 11 months.

The Small Tortoiseshell (*Aglais urticae*, 86, underside 87) has markings similar to those of the Large Tortoiseshell, but is placed in a separate genus on account of structural differences. It is common all over Europe and Asia, including Britain, and has two broods a year over most of its range. It often hibernates in houses, and the caterpillar feeds on stinging nettles.

88

89

88, 89 *Lesser Purple Emperor (Apatura ilia), larva and pupa* The larva has been photographed in front view to show the two remarkable horns that adorn the head. The resemblance of the pupa (89), both in colour and form, to a torn or curled leaf is very obvious and must serve to protect it from the eyes of searching birds.

The first complete and well illustrated book on British butterflies was written by Richard South and appeared in 1906. His description of the metamorphosis of the Purple Emperor (*Apatura iris*, see 54, 55) applies almost equally well to that of the species shown on this page, and we quote it here:

'The egg may be looked for in August on the upper surface of the leaf of the sallow (*Salix caprea*). It is pale olive green in colour and cylindrical in shape, the height from base to top being about equal to the width through from side to side. It has about fourteen ribs.

The caterpillar in October, just before hibernation, is dingy green roughened with numerous whitish warts from which arise short bristles, some of the latter appearing to be tinged with reddish, and those along the sides longer than those on the upper part of the body; the straight lines along the back and the oblique ones on the sides are yellowish. The head and the two horn-like projections, reminding one of the horns of a slug, are reddish-grey and covered with warts and bristles. The anal points (tails), which lie close together, are tinged with reddish. It should be mentioned here that on emerging from the egg the young caterpillar is without horns; these are not developed until the first skin is thrown off, which event happens from eight to twelve days after hatching.

The full-grown caterpillar is green, merging into yellow towards the anal points; the oblique stripes on the sides are yellowish tinged with reddish. The individual depicted on the plate [in South's book] took up a position for change to the chrysalis on June 6. It spun a mat of silk to the under side of the sallow leaf and the next day was found suspended by the claspers, which were grasping the silken mat. On the fourth day the chrysalis was fully developed, and from this a male butterfly emerged on June 24, an unusually early date.

The chrysalis is whitish, more or less tinged with green, but having the oblique lines on the sides whitish; the veins of the wings also show up whitish.'

90

90 *Knapweed Fritillary (Melitaea phoebe)* This is one of the 'checkerspot' types of Fritillary, lacking metallic spots on the underside. It is shown here shortly after eclosion, or hatching, from the chrysalis, the ruptured and empty shell of which is hanging from the twig just above the butterfly. It is suspended head-downwards by a single attachment at the tail, the mode of pupation found in all Nymphalid butterflies.

For a short time after it has left the chrysalis and expanded its wings a butterfly remains unable to fly because the wings have not yet dried and hardened. This is the ideal time to obtain photographs of living butterflies because there is no danger of the specimen flying away, and because its condition is perfect to the last hair and scale.

91

91, 92 *Nymphalid butterflies* These are photographs taken in life of two species which we have already shown mounted. The White Admiral (*Limenitis camilla*, 91) is also shown on plates 59 and 60. It is a lover of woodland flowers, and the one seen here is settled on a flower of one of the umbelliferous plants; it is perhaps most frequently seen taking nectar from the flowers of bramble or blackberry. The White Admiral is a shy butterfly, and the photographer's best hope of getting a picture of it in the wild is to stalk it carefully when it is feeding.

The Queen of Spain Fritillary (*Issoria lathonia*, 92) is seen on plates 61, 62, and the beautifully ornamented underside will immediately be recognised in this additional picture. The origin of its English name is obscure, but its rich adornment of silver spots may have prompted a reference to the nation which, in the eighteenth century, owned the greatest silver mines in the world.

93

94

93, 94 *Fritillaries* The two Fritillaries on this page are easily confused and hardly to be distinguished in flight. Plate 93 shows a Dark Green Fritillary (*Mesoacidalia aglaja*) settled close to a Common Blue (*Polyommatus icarus*). The other photograph (94) is of a High Brown Fritillary (*Fabriciana adippe*). In this species, on the underside of the hind-wing outside the main row of silver spots, there is a row of light-centred red-brown spots, which are wholly lacking in the other species. Also this is a woodland butterfly, while the Dark Green Fritillary flies over moorland and similar open country.

Both these Fritillaries extend right across Europe and Asia, from Britain in the west to Japan at the faraway eastern limit of Eurasia.

95

95 *Red Admiral (Vanessa atalanta)* Like many butterflies with the upper side brightly coloured (see 74) the Red Admiral has the underside, especially of the hind-wings, coloured and patterned for concealment when the insect is at rest. The photograph shows well the way in which almost the whole of the fore-wing is concealed by the hind-wing when the butterfly is settled. Another obvious characteristic is the essentially four-legged or quadruped nature of the Nymphalid butterflies; the small, stunted forelegs are hardly to be seen.

The Red Admiral's specific name is taken from Greek mythology, and refers to its beauty combined with its swift and graceful flight. Atalanta was an athletic princess who insisted that all suitors for her hand should test their manhood by running a race with her, but she easily outran them all and remained a maiden. At last one of them sought the help of Aphrodite the goddess of love and obtained from her three golden apples. Having challenged Atalanta to run against him, he sprinted ahead of her and dropped the apples, one by one, in her path. Her concern to stop and pick them up lost her the race and gained him a bride. In the eighteenth century butterfly book 'The Aurelian' this butterfly is called the 'Red Admirable', and the name by which we know it is a corruption of this and has no original reference to any naval officer.

The chrysalis of this, and many other nymphalid butterflies, is ornamented with shining metallic spots; the word 'chrysalis', derived from the Greek *chrysos* meaning gold, refers to this, though it is now used as an alternative to the term 'pupa' and extended to all the Lepidoptera and some other insects.

The word 'Aurelian', which Moses Harris used as the title for his book, is the name by which the early collectors of butterflies and moths called themselves; it has a similar derivation (Latin: *aurum*, gold) and also refers to the metallic ornamentation of some of the Nymphalid pupae. Why these people used such an unimportant and restricted characteristic of the insects they collected, to designate themselves, is hard to explain. It seems that these early collectors were extremely fanciful, and also that they liked to make a mystery of their pursuit, possibly because they were generally regarded as quite mad. This is literally true, for there is a well attested story of an eighteenth century lady, named Glanville, who collected butterflies and was at the same time fairly wealthy. When she died her will was contested by some relatives, who were disappointed in their expectations, on the grounds that because she collected butterflies she must have been insane; we are glad to say that when the matter was brought to court, judgment was given against them. This entertaining fragment of history is commemorated by the English name of the species *Melitaea cinxia*, the Glanville Fritillary.

96

97

98

99

100

96–100 *American butterflies* The butterflies of North America have much in common with those of Europe and temperate Asia, and we have already mentioned a number of species which are found in both regions. In the southern States many of the butterflies have a 'Neotropical' aspect, that is to say their evolutionary ancestors were probably South American. In addition, of course, there are numerous species found only on the North American continent.

The Tiger Swallowtail (*Papilio glaucus*, 96, 97) is a familiar butterfly of the United States, ranging from north of the border, in Canada, into the southern States. At first sight it looks as if we had made a mistake; how can 96 and 97 be the same species? Yet they are, and it is not a matter of the sexes being different. Over the whole region males and females more or less like 96 are found, but in the south two forms of the female occur, the normal one and the curious blackish form shown at 97. The larva feeds on leaves of various trees.

The Great Spangled Fritillary (*Speyeria cybele*, 98) is widely distributed and resembles some of the European Fritillaries in appearance. The Dog Face (*Colias cesonia*, 99), named for the outline of its black markings, is an abnormal Clouded Yellow and is found mainly in the south.

The lovely butterflies of the genus *Callicore* are characteristic of tropical America and come north only as far as Mexico. There are numerous species, many of them hard to distinguish from each other; the pattern of the underside, shown at 100, is their most striking feature.

101

102

103

101–103 *Vanessid butterflies* We have seen both these butterflies before on pages 40 and 43. The Painted Lady (*Vanessa cardui*, 101) has the underside lighter in colour than the other Vanessid butterflies, probably to harmonise with the withered and bleached vegetation of the semi-deserts which are its main breeding grounds.

As already noted (84, 85) the Peacock (*Inachis io*, 102, 103) feeds as a larva on nettle, and 102 shows a fully grown caterpillar on its food plant. The possession of three pairs of jointed 'true' legs is the hallmark of the insects. The larvae of butterflies and moths have these legs; in this caterpillar they are black, and can be seen at the left of the picture just behind the insect's head. Larvae of Lepidoptera have, in addition, a number of claspers or 'prolegs', usually five pairs. These can be plainly seen here because the anterior four pairs are red in colour, the hindmost one being black. The prolegs are soft and unjointed, but are provided with tiny chitinous hooks which enable the caterpillar to cling tightly to a stem or leaf.

The Peacock butterfly likes to hibernate in a dark place, such as a hollow tree or the attic of a house. In such situations its very dark underside coloration (103) gives it protective concealment.

104

105

106

107

108

104–108 *Satyrid butterflies* With this photograph of a Peacock butterfly displaying the beautiful upperside of its wings (104) we end our series of Nymphalid butterflies and make a start on another family, the Satyridae, often collectively known as the Browns, Heaths and Ringlets. They are almost always coloured in various shades of brown and dull yellow and have eye-like spots on the wings. One or more of the veins of the fore-wings are swollen at the base, as seen in 101 and 102. As in the Nymphalidae, the forelegs are stunted and brush-like. The larvae live on grasses of various kinds but are seldom seen, as they feed at night and shelter at the roots of the grass during the day. The pupae may be suspended by the tail or they may lie on the ground in a slight cocoon.

The Wall Brown (*Lasiommata megera*, 105) flies in open country, often in gardens, and has the habit of basking on rocks and walls in the sunshine. The specimen shown is a male and has on the fore-wings a pair of dark bars of special scales called androconia; these are scented and play a part in courtship. In the female they are wholly lacking. The Wall Brown is found in Europe and North Africa and extends into Asia Minor as far as Iran.

The Speckled Wood (*Pararge aegeria*, 106) is common all over Europe, where it occurs in two distinct forms. The one with yellow markings flies in the southern part of the region, but in northern and central Europe, including the British Isles, the pale markings are white. This species has the unusual habit of passing the winter either as a larva or a pupa. Most butterflies invariably winter in one particular stage of their development.

The False Ringlet (*Coenonympha oedippus*, 107, 108) is plain and drab on its upper side, but the underside (108) is very pretty. It is local in distribution and flies in marshy areas in the lowlands in Europe and temperate Asia.

109 110

111 112

113

114

115

109–115 *Satyridae* The species of *Erebia* (109, 110) are perhaps the most interesting of all the Satyrid butterflies. They are highly characteristic of the mountains of Europe, Asia and northern and western North America, occurring near sea level only in the northern and arctic regions. As mountainous districts tend to be isolated from each other by tracts of intervening lowland, numerous distinct species of *Erebia* have evolved on the various mountain ranges. Some are wholly restricted to the Alps, or even to certain parts of them; for example, *Erebia christi* occurs only in the southern Alps of Switzerland. Others are restricted to the Pyrenees, and the Black Ringlet (*Erebia melas*) is confined to the mountains of the Balkans. Several species are found only in the far north and arctic regions.

The Scotch Argus (*Erebia aethiops*, 109) has a fairly wide distribution and is one of the two species that occurs in Britain, where it flies on hillsides or along the borders of woods in damp localities in Scotland. It does not extend far up the mountains. It is widespread in Europe, and eastward to the Caucasus and Ural mountains. The larva feeds on grasses, and when ready to pupate it burrows into a tuft of grass and makes a hollow chamber lined with silk which hides and protects the chrysalis. The Woodland Ringlet (*Erebia medusa*, 110) has a rather similar distribution in Europe but does not extend to Britain. In its northern range it is a lowland butterfly, but may ascend to 4,000 feet further south.

Nearly all the species of *Erebia* are dark in colour; this is probably because their habitat is mostly cold and windy, and they benefit by absorbing as much heat as possible from the sun. As is well known, dark surfaces absorb radiant heat far more readily than light-coloured ones.

The Dusky Meadow Brown (*Hyponephele lycaon*, 111, 112) belongs to a group of Satyrid butterflies in which the sexes have a very different appearance. The male, on the left, is dusky brown with a single eye-spot on the fore-wing. The rather larger female has two conspicuous spots on each fore-wing, set in an area of pale yellow, and the whole coloration is much paler than in the male. The Dusky Meadow Brown has a rather southern distribution in Europe, and flies in dry, rocky places at sea level and on the lower mountain slopes.

The Meadow Brown (*Maniola jurtina*, male 113, female 114) also exhibits sexual dimorphism; the female has a single white-centred black spot in an orange area, near the apex of each fore-wing. The size of this orange patch is subject to a great deal of variation. The butterfly is very common in grassy places all over temperate Europe including Britain.

The Hedge Brown or Gatekeeper (*Pyronia tithonus*, 115, female) is a smaller butterfly, also common in Britain and continental Europe. The male has a conspicuous oblique black bar of specialised scales on each fore-wing. The Gatekeeper flies along country lanes and around the borders of woods, and is particularly fond of bramble flowers.

116

116 *The Monarch Butterfly (Danaus plexippus)* This American species is the most celebrated of all the migratory butterflies. In the United States and Canada its movements have been accurately traced by the method of marking and releasing butterflies, some of which are subsequently recaptured. Details of the marking will exactly identify the individual, which can then be presumed to have flown from its point of release to that of its recapture. In North America it is seen in summer all over the United States and southern Canada. In autumn the Monarchs in the north begin to move south, travelling in groups or swarms which fly by day and roost at night in trees. When they get as far as southern California and Florida they settle in closely packed crowds on trees and pass the winter in a state of semi-hibernation, occasionally flying round on sunny days. The same trees are used as resting places year after year, and in some places the butterflies are protected by law from disturbance. In spring they fly northward, the females laying eggs as they go, but this northward movement is less easy to observe as they go singly, frequently stopping on the way. Two or three generations are born through in the summer, the butterflies of each generation continuing to press northward. Then in the autumn they turn south again. On the southward migration an individual Monarch may travel 2,000 miles.

Their migrations take them outside America and they sometimes cross the Atlantic, carried by strong winds, and reach Britain and Europe. The Monarch has become established in Hawaii, Australia, New Zealand and some of the islands of the Malay archipelago, and also in the Canary Islands in the Atlantic. The widespread establishment by man of the milkweeds on which they feed is the reason for this recent extension of their range.

117 118

119 120

117–120 *Satyridae* On this page are two of the largest and finest Satyrid butterflies found in Europe, and they both belong to the group known as 'Graylings'. The Woodland Grayling (*Hipparchia fagi*, 117, male; 118, female) has a wing-span of two and a half inches and the female, as in many Satyrids, is more boldly and brightly marked than the male. It flies in central and southern Europe, eastwards into Russia, and is mainly a butterfly of the lowlands. It often rests on tree trunks and, as its name implies, is more a butterfly of wooded country than the other Graylings. The Rock Grayling (*Hipparchia alcyone*) often flies together with the Woodland Grayling, but frequents rough, rocky country and ranges further up the mountains. The two species are very alike, but the Rock Grayling is a little smaller and the pale markings more inclined to yellowish.

The Great Banded Grayling (*Brintesia circe*, 119, 120) is larger than the Woodland Grayling and the most handsome of all the European Satyrids. Here there is little difference between the sexes, both having the upper side almost black with a broad white broken band crossing the fore- and hind-wings. The underside (120) is patterned to give protection on a background of bark, rock or open ground where the butterflies are most likely to settle.

121

122

121–123 *Satyridae* The two butterflies on this page have already been portrayed on previous pages. The Wall Brown (*Lasiommata megera*, 121 and 105) is feeding at a flower, and is moving actively about and making no attempt to hide. Its fore-wing is raised and the conspicuous eye-spot near its tip is displayed. It is probable that this spot, and the other eye-spots on the underside, play a part in protecting the butterfly from the consequences of attack by birds. Because the spots are the most conspicuous and striking features visible when the insect is not flying, a bird is more likely to peck at them than at any other part of the butterfly. If it does so part of the wing will probably tear away and the butterfly will escape, as loss of a small area of wing does not affect its ability to fly. This theory that the spots serve to divert the attention of predators from the insect's body receives support from the fact that butterflies are sometimes captured with the wings torn just where the spots are located, or with clear marks of birds' beaks on them.

The Woodland Grayling (*Hipparchia fagi*, 117, 118, 122) also has an eye-spot near the fore-wing apex on the underside. The butterfly photographed, however, was at rest, not moving about. It has therefore dropped the fore-wings down between the hind-wings, concealing the eye-spot and enabling its natural camouflage to give it maximum protection.

The Marbled White (*Melanargia galathea*, 123) is the commonest member of a group of Satyrid butterflies that depart from the usual sombre coloration of the family and display a bold checkered pattern of black and white on the upper side, and on the underside the delicate network of lines and eye-spots that the photograph shows. Several rather similar species are found in Europe and western Asia, but the genus does not extend to America.

123

124

124 *Cecropia moth (Hyalophora cecropia)* The family Saturniidae is poorly represented in Europe and we show only one of the species, *Saturnia pyri* (14, 15). In North America more than sixty species are found, some very large and fine. The one shown here, also known as the Robin moth, is one of the biggest of them, and is equalled in size by the Polyphemus moth (*Antheraea polyphemus*), which is tan coloured with a translucent eye-like spot on each hind-wing.

The greatest variety of Saturniids is found in the tropics, and the largest species of all occur in the Oriental and Australian tropical regions. There we find the Giant Atlas moth (*Attacus atlas*) and the enormous Hercules moth (*Coscinoscera hercules*). This is the largest of all Lepidoptera and the total area of its wings may exceed 40 square inches. It lives in the rain forests of New Guinea and northern Australia.

Many of these large Saturniids are easy to breed in captivity and their larvae will feed on a variety of trees and shrubs that are generally available. The Cecropia moth can be bred on willow, hawthorn or cherry, the Polyphemus moth on oak and the Giant Atlas on privet. There are dealers in many countries who maintain stocks of these and other large and spectacular moths and sell eggs and living larvae or pupae with instructions for rearing them.

125 126

127 128

125–128 *Tropical butterflies* Very little observation of their habits is needed to show us that butterflies are creatures of the sun, and they could be expected to flourish most in those regions where hot, brilliant sunshine is a dominant feature of the climate. Since almost all of them feed as larvae on the leaves of plants and trees, abundance and diversity of vegetation is another factor likely to lead to a corresponding diversity and abundance of butterflies. Thus it is that the tropical rain forest far exceeds any other type of environment in the splendour and variety of its butterflies. This is true of the rain forest of all the continents where it is found, Central and South America, Africa, Asia and the extreme north of the Australian region.

On this page three of these regions are represented: *Adelpha iphicla* (125) and *Chlorippe seraphina* (128) are South American; *Papilio nireus* (126) is an African Swallowtail; *Papilio demoleus* (127) is a common Swallowtail of tropical Asia. The last species may be seen in the forest, but is a familiar garden butterfly as its larva feeds on lime and other kinds of Citrus bushes, which are widely cultivated for their fruit.

129 130

131 132

129–136 *Tropical butterflies* Plate 129 shows *Graphium policenes*, an African member of a group of butterflies known as Sword-tail Swallowtails on account of the slender curved 'tails' on the hind-wing resembling a sword blade. They are found in Africa, tropical Asia and Australia, and may sometimes be seen in numbers drinking on patches of moist sand beside rivers. *Papilio paris* (130) is a Swallowtail with a wide distribution from India to Indonesia.

The greatest variety of tropical butterflies is to be found in the rain forests of Central and South America, especially that of the Amazon Basin in Brazil and Peru. Among the most vividly coloured of them are the species of *Perisama*, one of which is shown at 131. *Smyrna blomfildia* (132) has a wide distribution in tropical America from Mexico southward to Paraguay and northern Argentina.

Although South America has the richest fauna of tropical butterflies, these do not include the largest species. The world's biggest butterflies are the Birdwing Swallowtails (*Ornithoptera*) of New Guinea and many of the smaller islands of the region, and tropical Australia. The males are large and brilliantly coloured; the females are always brown with paler markings but even larger. The female of *Ornithoptera priamus* is shown at 133. It may reach eight inches in wing-span, exceeded by that of *O. alexandrae* which attains over ten inches and is the largest butterfly in the world. *Danaus chrysippus* (134) is found in both Asia and Africa and is closely related to the Monarch (116). Both belong to a family of butterflies that are distasteful and poisonous to birds, which quickly learn to recognize them and leave them alone. *Panacea prola* (135) and *Phoebis cipris* (136) are both South American species.

133

134

135

136